Lecture Notes in Mathematics 1469

Editors:
A. Dold, Heidelberg
B. Eckmann, Zürich
F. Takens, Groningen

J. Lindenstrauss   V. D. Milman (Eds.)

Geometric Aspects
of Functional Analysis

Israel Seminar (GAFA) 1989-90

Springer-Verlag

Berlin Heidelberg New York
London Paris Tokyo
Hong Kong Barcelona
Budapest

Editors

Joram Lindenstrauss
Hebrew University of Jerusalem
Givat Ram, 91 904 Jerusalem, Israel

Vitali D. Milman
School of Mathematical Science
The Raymond and Beverly Sackler Faculty of Exact Sciences
Tel Aviv University
Ramat Aviv, 69 978 Tel Aviv, Israel

Mathematics Subject Classification (1985):

Primary: 11M99, 42B15, 52A20, 52A40, 52A45, 54H20, 81E99
Secondary: 46B10, 46B25

ISBN 3-540-54024-5 Springer-Verlag Berlin Heidelberg New York
ISBN 0-387-54024-5 Springer-Verlag New York Berlin Heidelberg

© Springer-Verlag Berlin Heidelberg 1991
Printed in Germany

Printing and binding: Druckhaus Beltz, Hemsbach/Bergstr.
2146/3140-543210 - Printed on acid-free paper

FOREWORD

This is the fifth published volume of the proceedings of the Israel Seminar on Geometric Aspects of Functional Analysis (GAFA). The previous volumes are

1983-84   published privately by Tel Aviv University

1985-86   Springer Lecture Notes, Vol. 1267

1986-87   Springer Lecture Notes, Vol. 1317

1987-88   Springer Lecture Notes, Vol. 1376

As usual, the large majority of papers in this volume are original research papers; the others are surveys which include a considerable proportion of original research material. They are arranged, on the whole, in the order in which they were presented at the seminar.

We are grateful to Mrs. M. Hercberg, without whose help this volume could not have been prepared.

Joram Lindenstrauss, Vitali Milman

1989–1990

GAFA 1989-1990

List of Seminar Talks

3 November 1989   1. L. CARLESON (Royal Institute of Technology, Stockholm) Chaos and order in elementary dynamical systems.
2. V. MILMAN (Tel Aviv University) Some geometric duality relations.

10 November 1989   1. L. CARLESON The Hénon map and the standard map.
2. G. SCHECHTMAN (Weizmann Institute, Rehovot) Complemented subspaces of ℓ_p^n (joint work with W.B. Johnson).

26 November 1989   1. G. SCHECHTMAN What is left of the L_p^n ball after subtracting a multiple of the L_q^n ball? (joint work with J. Zinn).
2. D. BURKHOLDER (University of Illinois) Some explorations in martingale theory and Banach spaces.

10 December 1989   Y. FRIEDMAN (Jerusalem College of Technology)   1. An algebraic category in non-commutative geometry   2. Classification of Cartan domains. How exceptional domains occur.

22 December 1989   1. Y.G. SINAI (Landau Institute of Theoretical Physics, Moscow) Mathematical problems in the theory of quantum chaos.
2. H. FURSTENBERG (Hebrew University) Non-conventional ergodic averages and nilpotent groups.

7 January 1990   1. A. REZNIKOV (Tel Aviv University) Isoperimetric inequalities for simplicies.
2. Y. STERNFELD (Haifa University) Extreme points of convex bodies in euclidean spaces.

21 January 1990   J. LINDENSTRAUSS (Hebrew University) Characterizing c_0, ℓ_1 and $c_0 \oplus \ell_1$ by 2 absolutely summing operators (after Rudelson).

4 March 1990   1. A. REZNIKOV Norms on tensor products and characterizations of Hilbert spaces.
2. H. KÖNIG Some estimates for entropy numbers (after M. Defant and M. Junge).

16 March 1990   1. M. KRASNOSELSKY (Moscow) A non-conventional fixed point theorem.
2. N. TOMCZAK-JAEGERMANN (University of Alberta) Non-conventional Hilbert spaces with unconditional bases (joint work with N. Nielsen).

30 March 1990   1. N. KRUPNIK (Rehovot) On the norm of polynomials of two projections in Hilbert space (joint work with I.A. Feldman and A. Markus).
2. V. MILMAN On recent results in Local Theory (on results of Talagrand, Kashin and others)

20 April 1990   1. G. KALAI (Hebrew University) Two combinatorial isoperimetric theorems.
2. P. MÜLLER (Linz University and Weizmann Institute) On permutation of the Haar system.
3. V. MILMAN Information on new results on finite metric spaces (after J. Matoušek).

27 April 1990   1. G. MARGULIS (Institute on Problems of Information Transmission, Moscow) On a simplified proof of the Oppenheim conjecture.

2. J. BOURGAIN (IHES, France) Volumes of sections of convex bodies in R^n.

6 May 1990   1. A. PELCZYNSKI (Polish Academy of Sciences, Warsaw) Some remarks on John's ellipsoid.

2. H. GLUSKIN (Tel Aviv University) A local version of Menshov's Theorem (after Kašin).

18 May 1990   1. W.B. JOHNSON (Texas A&M University) Characterizing weak Hilbert spaces by approximation properties (joint work with G. Pisier).

2. J. BOURGAIN Distribution of polynomials on convex sets.

3. A. PELCZYNSKI Parallelepipeds of minimal volume containing a symmetric convex body (joint work with S.J. Szarek).

3 June 1990   1. Y. BENYAMINI (Technion, Haifa) Characterization of harmonic and holomorphic functions by the mean value property (joint work with Y. Weit).

2. J. LINDENSTRAUSS Covering sets in R^n by balls of the same diameter (joint work with J. Bourgain).

3. S. REISNER (Haifa University) Characterizing logconcave and affinely rotation invariant measures by the location of the centroids of their sections (joint work with M. Meyer).

TABLE OF CONTENTS

STOCHASTIC MODELS OF SOME DYNAMICAL SYSTEMS

Lennart Carleson*

K.T.H.
Stockholm, Sweden
and
U.C.L.A.
Los Angeles, USA

The purpose of this informal paper is two-fold: (a) to present a survey of the basic ideas in the papers [1] and [2] and in this way make these more easily accessible and (b) to exhibit what I believe is the basic mechanism in a large class of non-uniformly expanding systems.

We shall discuss three examples of mappings

(A) $f : x \to 1 - ax^2$, $x \in (-1,1)$, $a_0 < a < 2$

(B) the Hénon map:

$$T : \begin{cases} x' = 1 - ax^2 + y \\ y' = bx \end{cases} \qquad \begin{array}{l} a_0 < a < 2 \,,\, 0 < b < b_0 \\ (x,y) \in \mathcal{D} \,,\, T(\mathcal{D}) \subset \mathcal{D} \end{array}$$

and an area preserving map:

(C)

$$x' = a\varphi(x) - y$$

$$(\mathrm{mod}\, 1)$$

$$y' = x$$

a is large and $\varphi(x)$ smooth e.g. $= \sin 2\pi x$.

For example (A) it is known that the mapping has chaotic behavior (e.g. in the sense that there exists an absolutely continuous invariant measure) if the Liaponov exponent for $x = 1$ is positive. This means that if $|x_1| < 1, \ldots, x_n = f(x_{n-1})$ and

$$\mathcal{D}_n^* = \prod_1^n 2a|x_\nu|$$

* This work was carried out at the Institute for Advanced Study, Princeton, during Spring 1989.

then for almost all x_1

$$\lambda = \lim \frac{\log \mathcal{D}_n^*}{n}$$

exists > 0. Also for the other mappings we shall concentrate on proving positive Liaponov exponents for certain critical choices of (x, y) without discussing further the (still incomplete) theory how this implies e.g. existence of smooth invariant measures. This extension of Pesin's theory seems to be possible along well understood lines.

1. The 1-dimensional case

A. A model. We shall define a sequence $\mathcal{D}_\nu(\omega)$, $\mathcal{D}_0 \equiv 1$, of random functions and a sequence of stopping times $\nu_j(\omega)$, $\nu_1 \equiv 1$. We start from a fixed number $a > 0$ and do the following construction.

If ν_j is a stopping time we choose t_j, $0 < t_j < 1$, at random with uniform distribution, and independent of the past. We then define

$$\mathcal{D}_{\nu_j + \mu} = \mathcal{D}_{\nu_j - 1}(2at_j) \cdot \mathcal{D}_\mu \tag{1.1}$$

for $\mu = 0, 1, \ldots, \mu_j - 1$ where $\mu_j \geq 1$ is the smallest integer for which

$$t_j^2 \mathcal{D}_{\mu_j} > 1 . \tag{1.2}$$

If no μ_j exists we use (1.1) for all μ.

Theorem 1. *If $a > \frac{1}{2}e$ the Liaponov exponent*

$$\lim \frac{\log \mathcal{D}_\nu(\omega)}{\nu} \tag{1.3}$$

is positive with positive probability.

The proof is an exercise. We first choose $t_1, \ldots, t_N$ close to 1 and in this way obtain

$$\mathcal{D}_\mu > (2a - \varepsilon)^\mu , \qquad \mu = 1, \ldots, N . \tag{1.4}$$

Once we know that these $\mathcal{D}_\mu$ are large the only way in which $\mathcal{D}_\nu$ would fail to grow exponentially would be by persistent small choices of t_j. Since however the mean value

$$\int_0^1 \log(2at)dt > 0$$

this is exponentially unlikely.

Remark. It is a (mildly) interesting problem to decide what is the best condition $a > a_0$ for the validity of Theorem 1.

B. We now study mapping $f(x; a) = 1 - ax^2$. We define

$$\mathcal{D}_\nu(a) = \frac{d}{dx} f^\nu(1; a) .$$

Theorem 2 (Jakobson[3]) [1]. *For a set of positive measures of the parameter a, $0 < a < 2$, the Liaponov exponent λ is positive.*

Let us now try to see how Theorem 2 is related to Theorem 1. We first observe that

$$\mathcal{D}_\nu(a) = \prod_{j=0}^{\nu-1} (-2a\xi_j) \tag{1.5}$$

where

$$\xi_j = f^j(1; a) .$$

We decide to study the problem for $a \in \Delta = (A, B)$ with A, B close to 2.

The first part of the proofs of Theorem 1 and Theorem 2 are now quite similar. This is obvious but something much stronger holds.

Lemma. *Let $\delta > 0$ be given. Let $\{x_j\}_1^N$ be an orbit with $|x_j| > \delta$ and assume $|x_{N+1}| < \delta$. Then if $a > 2 - \varepsilon(\delta, N)$*

$$\prod_1^N |2ax_j| > (2 - \delta)^N .$$

The proof of the lemma is easy since $1 - ax^2$ is close to $1 - 2x^2$ and $1 - 2x^2$ is conjugate to $1 - 2|x|$.

The second fact is that for iterations $f^n(x; a)$ derivatives with respect to the parameter a and with respect to the variable x have pointwise bounded ratio from above and from below if the x-derivative $\mathcal{D}_\nu$ satisfies

$$\sum |\mathcal{D}_\nu|^{-1} < M$$

and if we have suitably large $\mathcal{D}_\nu$ for some initial values of ν. — This in particular holds in the case of exponential increase.

Let us now consider our sequence (1.5). For $\nu \leq N$ we have increase as $(4 - \delta)^\nu$ $(a > a_0)$. The parameter interval Δ is mapped to $I_\nu = f^\nu(1; \Delta)$ expanding at the same rate and does not fold back on itself until $|I_\nu| \sim 1$ if Δ is chosen carefully. This is our first stopping time. The scale is uniform in the mapping $\Delta \to I_{\nu-1}$.

We now consider $a \in \Delta_1$ so that $e^{-k-1} \leq |f^\nu(1; \Delta_1)| < e^{-k}$. We obtain the derivative

$$\mathcal{D}_\nu \sim \mathcal{D}_{\nu-1} \cdot 2ae^{-k} \sim \mathcal{D}_{\nu-1} 2a\xi_\nu$$

for these parameter values. In the next iteration the x-interval $(0, e^{-k})$ is mapped into $(1 - 4e^{-2k}, 1)$ of length $\sim \xi_\nu^2$ and

$$|\xi_{\nu+j} - \xi_j| \leq C|\mathcal{D}_j|\xi_\nu^2$$

and hence

$$\frac{1}{C} \leq \frac{\prod_{j=0}^m |2a\xi_{\nu+j}|}{\prod_1^m |2a\xi_j|} \leq C$$

as long as (a)$\mathcal{D}_j$ increases exponentially, (b) $|\xi_j|$ are not small compared to $|\mathcal{D}_j|\xi_\nu^2$ and finally (c) $|\mathcal{D}_j|\xi_\nu^2 < 1$. If we decide to ignore (b) we see that

$$\mathcal{D}_{\nu+j} \sim \mathcal{D}_{\nu-1} 2a\xi_\nu \mathcal{D}_j$$

as prescribed in the model and with the stopping rule (1.2). When the stopping rule applies we are again in the same situation with an interval $f^n(1; \Delta_1)$ of length 1 and with a uniform mapping from Δ_1 to this long interval.

The only difficulty is to prove that we may ignore (b) and this is done by avoiding small early returns, i.e. requiring for example $|\xi_j| \geq e^{-\alpha j}$ (α small fixed). We must then also face the difficulty that we may get no parameters left and also that returns $\xi_j \sim e^{-\alpha j}$ may destroy the exponential increase (see [2]).

C.   The model can easily be generalized to a finite set of choices of functions $\mathcal{D}_\nu(\omega)$. Let us for simplicity consider two functions denoted $\mathcal{D}_\nu'$ and $\mathcal{D}_\nu''$ and two associated sequences of stopping times, denoted ν_j' and v_j''. We simply change the rule (1.1) by writing

$$\mathcal{D}_{\nu_j'+\mu}' = \mathcal{D}_{\nu_j-1}'(2at_j') \cdot \begin{cases} \mathcal{D}_\mu' \\ \text{or} \\ \mathcal{D}_\mu'' \end{cases} \tag{1.6}$$

$\mu = 0$, $\mu_j' - 1$ with unchanged stopping rule (1.2). We make a similar definition for $\mathcal{D}''$. The choice of $\mathcal{D}'$ or $\mathcal{D}''$ in (1.6) should be the same for $\mu < \mu_j'$ in realistic models but we are allowed to use *any* rule which is measurable on the past (i.e. times $< \nu'_j$). It is still easy to prove that we have exponential increase with positive probability.

The indicated model is a model for the dynamics of a mapping $x \to f(x; a)$ where $f(x; a)$ is a non-degenerate family of functions f mapping $(-1, 1)$ into itself with a finite number of extreme points and one can rigorously prove the analogue of Theorem 2, provided we have suitable information on the early orbit.

A slightly different finite generalization is obtained as follows. Let $a_1, a_2, \ldots, a_n$ be independent choices of parameter values and consider the functions $f_i = 1 - a_i x^2$, $a_0 < a_i < 2$. Let the statespace be $(i; x)$ $i = 1, \ldots, n$; $-1 < x < 1$. Let $\pi_1(i)$ and $\pi_2(i)$ be permutations of $\{1, 2, \ldots, n\}$ and define the dynamics as follows

$$T : (i, x) \to \begin{cases} (\pi_1(i), f_i(x)), x > 0 \\ (\pi_2(i), f_i(x)), x < 0 \end{cases}$$

It is clear that the dynamics is again of the type (1.6) and one can rigorously prove that the mapping has positive Liaponov exponent for a set of positive measure of $(a_1, a_2, \ldots, a_n)$. This dynamics can be visualized as taking place on a fan of lines, each with its own dynamics

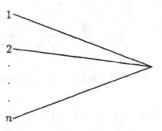

Figure 1.

and an arbitrary rule from each line to two others. This picture should be compared with the Hénon map that we shall now describe.

2. The Hénon map

A. The model. We shall here generalize the models of §1 by constructing random functions $\mathcal{D}_\nu(u;\omega)$. The domain of definition is a (thin) Cantor set E obtained by successively applying the rule $(b, 1 - 2b, b)$; b is related to the b-parameter in the Hénon map and is small > 0. As functions of $u \in E$ the functions have a continuity expressed by the statement that as long as $\mathcal{D}_j|u - u'| < 1$, $j = 1, 2, \ldots, m$, the functions $\mathcal{D}_j(u) \equiv \mathcal{D}_j(u')$, $j \leq m$.

We now make a formal definition of these ideas.

Let E be the Cantor set $(b, 1 - 2b, b)$ and let I denote an interval of length b^m used in the construction of E. $\mathcal{D}_\nu(u;\omega)$ are random functions defined on E, $\mathcal{D}_0 \equiv 1$, and each $\mathcal{D}_\nu$ is constant on some maximal I_ν. We also have stopping times $\nu_j(u)$, $\nu_1 \equiv 1$.

1. Let ν_j be a stopping time and $I_{\nu_j - 1}$ the associated interval of constancy. We choose t at random $0 < t < 1$, independent of all choices for times $< \nu_j$ and we take $v \in E$ arbitrarily (measurable on the past). We define for $\nu = \nu_j + \mu$

$$\mathcal{D}_\nu(u) = \mathcal{D}_{\nu_j-1}(u)t\mathcal{D}_\mu(v) , \qquad u \in I_{\nu-1} \tag{2.1}$$

and

$$I_\nu = I_{\nu-1}$$

for $\mu = 0, 1, \ldots$ as long as

$$\text{(a)} \qquad \mathcal{D}_\nu|I_{\nu-1}| < 1$$

and

$$\text{(b)} \qquad \mathcal{D}_\mu t^2 < 1 .$$

2. If (a) fails before (b) subdivide $I_{\nu-1}$ into two intervals of the next generation

$$I_{\nu-1} = I' \cup I'' \cup (\text{excluded middle interval}) .$$

Define $I' = I_\nu$ and $\nu_{j+1}(u) = \nu$ for $u \in I'$ and go back to rule 1 for $u \in I_\nu$. For $u \in I''$ define $I_\nu = I''$ and continue at (2.1) without changing t and v.

3. If (b) fails before (a) let $\nu = \nu_j + \mu$ be the definition of ν_{j+1}, $u \in I_{\nu_j-1}$, and go back to rule 1.

Theorem 3. *With $a > \frac{1}{2}e$ there exists $b_0 > 0$ so that if $b < b_0$ for some $\lambda > 0$*

$$\mathcal{D}_\nu(u;\omega) \geq c^{\lambda\nu} \qquad \nu = 1, 2, \ldots, u \in E \tag{2.2}$$

on a set ω's of positive probability independent of the choice of functions v in rule 1.

The proof of Theorem 3 is an elaboration of the proof of Theorem 1. It is true that we have to consider an infinite number of points which eventually become independent. However, as long as $\mathcal{D}_\nu < b^{-n}$ we only have 2^n different functions. If we therefore obtain exponentially small probabilities for failure of (2.2) for $\nu > m$, we can exclude all these exceptional sets and still have positive probability. This is carried out in detail in [2], section 3.

B. The connection between the model 2A and the dynamics of the Hénon map

$$T: \quad \begin{cases} 1 - ax^2 + y \\ bx \end{cases} \quad a_0 < a < 2, \quad 0 < b < b_0$$

is somewhat complicated.

There is a fixed point F in $(x > 0, y > 0)$ with eigenvalues $\sim 2, \sim \frac{b}{2}$. Let W^u, W^s be the unstable and stable manifolds of F. There is a bounded domain $\mathcal{D} \supset W^u$ such that $T(\mathcal{D}) \subset \mathcal{D}$.

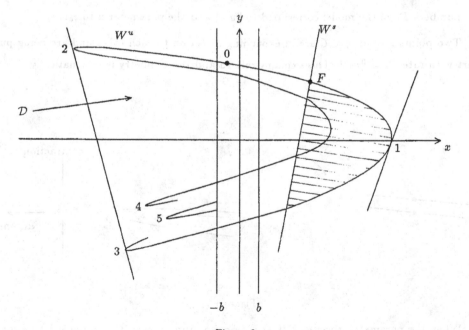

Figure 2.

The Jacobian of T is $= -b$. It is easy to see that e.g. the shaded area tends to $\overline{W^u}$ under forward iterations so that $\overline{W^u}$ contains the attractor for this domain. We shall therefore discuss only the dynamics of T on W^u.

It is intuitively plausible that there is a set $C \subset W^u$ in the strip $-b < x < b$ which has the property that tangents to W^u for points in C are mapped to the ultimate tangents at the folds of W^u. The slop of the tangent has to equal "the contractive direction"

$$q(x_0) = 2ax_0 + \cfrac{b}{2ax_1 + \cfrac{b}{2ax_2 + \ldots}} \tag{2.3}$$

as such a point $z_0 = (x_0, y_0)$, where the convergence of the continued fraction is very unstable. Note that for $a \sim 2\ b \sim 0$ (the slope at x_1) $= 2ax_1 + \frac{b}{2ax_2 + \ldots} \sim 4$ so the tangent is *not* vertical.

C is a Cantor type set with small ratio as $b \to 0$. Let us consider the forward iterates

$$w_\nu = \big(DT^\nu(z_0)\big)(0,1) \qquad z_0 \in C \ .$$

The numbers $\mathcal{D}_\nu$ of the model correspond to $\|w_\nu\|$ and the parameter u to z_0.

Two points z_0 and $z_0' \in C$ are in expanding direction to each other and are being pulled apart with rate $\mathcal{D}_\nu = \|w_\nu\|$. This explains rule 1a and the geometry is schematically

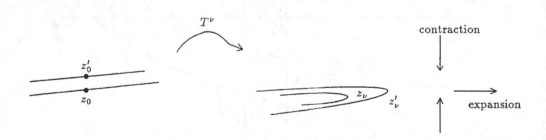

Figure 3.

When the parameter a varies, $T^n(z_0; a)$ traces a curve γ_n and this curve is a true copy of

W^u — the dotted line in the picture below – if we have exponential increase.

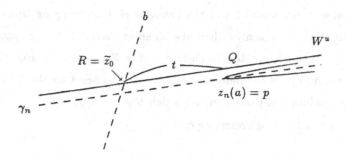

Figure 4.

The tangent direction of γ_n at P is essentially the same as the tangent direction of W^u at Q and essentially the same as the direction w_n. The reason is that all are approximately images of the same exponentially expanding matrix. The distance $PQ \sim b^n$. Let t be the distance RQ.

We split the vector w_n into its components for the two vectors $(0,1)$ and the unit vector e with direction q given in (2.3) with (x_0, y_0) replaced by (x_n, y_n):

$$w_n = D \cdot (0,1) + E \cdot e \,.$$

If $P \in C$ then by the definition of C, $D \sim 0$. If q follows its leading term $2ax_n$ in (2.3), the component D at (x_n, y_n) is proportional to the distance t to C (since the tangent direction changes very little).

This explains the factor t in (2.1). We should follow the orbit of the critical point $\tilde{z}_0$ (v in the model) as long as the points are close together and an analysis shows that this distance is again given by $t^2 \mathcal{D}_\mu$ as in rule 1b of the model.

It can now be proved [2] that all the above heuristics are indeed valid and the following theorem holds.

Theorem 4. *If $b < b_0$ there exists a set $E(b)$ of positive measure of parameter values a for which there is $z_0 \in W^u$ with $T^n(z_0)$ dense on W^u and $\|DT^n(z_0)\|$ has positive Liaponov exponent.*

From this information the mapping can no doubt be completely analysed but the details of this have not yet been carried out.

3.

The description in the model $2A$ of the behavior of T at stopping times is not a true description. First there is no analysis how the different sheets of W^u are approached – we resort to a worst case scenario. More importantly, if in Fig. 4 $RQ^2 < PQ$, the behavior is different and attractive cycles can indeed be created. In the proof for the Hénon map this is avoided simply by exclusion of parameters for which $RQ^2 < PQ$.

Let us now turn to the area preserving case

$$T : \begin{cases} a\varphi(x) - y \\ x \end{cases} \quad (\mathrm{mod}\,1)\,, \quad a \text{ large} \tag{3.1}$$

and let us for definiteness choose $\varphi(x) = \sin 2\pi x$. This mapping has strong expansion and contraction and a critical set C in two strips $|x - \frac{\pi}{2}, \frac{3\pi}{2}| < \delta$. We should think of C as a curve since W^u (at $(0,0)$ for example) is presumably everywhere dense for large a for sets of positive measure.

let us see how the model for the Hénon map needs to be modified.

I. The base set E should be an interval $= (0,1)$. We then use dyadic decomposition of E into intervals I when $\mathcal{D}_\nu |I| > 1$.

II. Contrary to the Hénon map, our mapping cannot contract. We should therefore study $\max(1, \mathcal{D}_\nu)$.

III. We can not afford in this case to use a worst case scenario in rule (2.1). There are good reasons for assuming that the level of return v is uniform in the vertical direction which would indicate that we should take v at random with probability measure $=$ Lebesgue measure on E.

IV. The parameter dependence is not exactly the same as space dependence. If, e.g. we consider the 1-dimensional iterations

$$x_{n+1}(a) = 1 - ax_n(a)^2$$

then

$$x'_{n+1} = (-2ax_n)x'_n - x_n^2 \,.$$

The first term behaves as space derivative which would now vanish if $x_n = 0$. In x'_{n+2} we would however obtain the constant contribution $-x_{n+1}^2 = -1$.

Similarly in 2 dimensions, if at time n with derivative $\mathcal{D}_n$ we return to $\mathcal{C}$ at ζ, the space expansion will be lost at time $n+m$, if the new derivative $\mathcal{D}_m(\zeta) = \mathcal{D}_n$. However, the parameter dependence is increased on scale $\mathcal{D}_n$. It is therefore natural to suppose that in a case like this we can have a stopping time.

All together this gives the following model.

Let $E = (0,1)$ and let I be dyadic subintervals of E. $\mathcal{D}_\nu(u;\omega)$ are random functions defined on E, $\mathcal{D}_0 \equiv 1$, and each $\mathcal{D}_\nu$ is constant on some maximal interval I. We also have stopping times $\nu_j(u)$, $\nu_1 \equiv 1$.

1. Let ν_j be a stopping time and I_{ν_j-1} the associated interval of constancy at time $\nu_j - 1$. We choose t at random on $0 < t < 1$ with uniform probability and v similarly in $E (= (0,1)$ also). We define for $\nu = \nu_j + \mu$

$$\mathcal{D}_\nu(u) = \left(t\mathcal{D}_\mu(v) + \frac{1}{\mathcal{D}_\mu(v)} \right) \mathcal{D}_{\nu_j-1}(u) \tag{3.2}$$

and

$$I_\nu = I_{\nu-1}$$

for $\mu = 0, 1 \ldots$ as long as

(a) $\mathcal{D}_\nu |I_{\nu-1}| < 1$

(b) $\mathcal{D}_\mu t^2 < 1$

(c) $\mathcal{D}_\mu < \mathcal{D}_{\nu_j-1}$.

2. If (a) fails first subdivide $I_{\nu-1}$ into two new dyadic intervals I' and I''. Define $I' = I_\nu$ and $\nu_{j+1}(u) = \nu$ for $u \in I'$ and go back to rule 1 for $u \in I_\nu$. For $u \in I''$ define $I_\nu = I''$ and consider again the alternatives (a),(b),(c).

3. If (b) fails first let $\nu = \nu_j + \mu$ be the definition of ν_{j+1}, $u \in I_{\nu-1}$, and go back to rule 1.

4. If (c) fails first again set $\nu_{j+1} = \nu_j + \mu$ and go back to rule 1.

Theorem 5. *With positive probability this model is exponentially increasing, e.g. in the sense*

$$\frac{1}{\nu} \int_0^1 \log \mathcal{D}_\nu(x;\omega)dx \geq \lambda > 0 .$$

This again is not so hard to prove. One needs here to keep track of estimates of the following type

$$m\{u \mid \mathcal{D}_j(u;\omega) \leq A\} \leq Ac^{-Mj}$$

where M is large when a is large. The above estimate clearly implies the theorem.

B. I have proof that T in (3.1) behaves as the model and in particular rule 4 can realistically only be expected to hold for one return. We then need (b) to apply before we can again use (c). An eventual proof must be considerably more complicated than the Hénon case. In particular, III is difficult to prove.

One test of the validity of the model concerns the existence of elliptic islands. Using the language of the model an elliptic fixpoint for T^k would be created as follows.

An interval $I \subset C$ is expanded up to time n at the rate $\mathcal{D}_n$. Hence $|I|\mathcal{D}_n < 1$.

At time n we have a stopping time and a choice t ($=$ distance to C). At time m after n the points return to C with bounded derivative. Hence by (3.2)

$$t\mathcal{D}_n(u)\mathcal{D}_m(v) \sim t\mathcal{D}_n^2 < 1 .$$

Now t measures the probability for this and $t \lesssim D_n^{-2}$. We also want our return interval at time $k = n + m$ which is again of length $|I| < \mathcal{D}_n^{-1}$ to return to the same place as I. The probability for this is $1/\mathcal{D}_n$ so that the total probability is $\mathcal{D}_n^{-3}$.

Conjecture. Let $T^k(z_0) = z_0$ be an elliptic fixpoint and let

$$M = \sup_{j < k} \left\| \mathcal{D}T^j(z_0) \right\| .$$

Then the size of the island around z_0 is $\sim M^{-3}$ and the island disappears if we move the parameter $\gg M^{-3}$.

For large a the density of the set of parameters for which any elliptic island exists, is small.

References

[1] M. Benedicks, L. Carleson, Ann. Math. 122 (1985).
[2] M. Benedicks, L. Carleson, The dynamics of the Hénon map. To appear in Ann. Math. (1990).
[3] M.V. Jakobson, Comm. Math. Phys. 81 (1981).

SOME APPLICATIONS OF DUALITY RELATIONS

V. Milman*

Raymond and Beverly Sackler
Faculty of Exact Sciences
Tel Aviv University
Tel Aviv, Israel

We demonstrate in this paper some deep relations between a finite dimensional normed space $X = (\mathsf{R}^n, \| \cdot \|)$ and its dual $X^* = (\mathsf{R}^n, \| \cdot \|^*)$. In geometric language, we will study relations between a convex, centrally symmetric, compact body $K \subset \mathsf{R}^n$ and its dual (also called polar) body

$$K^\circ = \left\{ x \in \mathsf{R}^n \mid (x, y) \leq 1 \quad \text{for every } y \in K \right\} .$$

Here (x, y) is the natural inner product in R^n equipped with the natural euclidean norm $|x|$. We always visualize K as the unit ball of the normed space X with the norm $\| \cdot \|_K$; then K° is identified with the unit ball of X^* with the norm $\| \cdot \|_{K^\circ} = (\| \cdot \|_K)^*$.

Results, which are presented here, show the "mystique" of duality: some strange regularization role of a dual pair of operations (Theorem 1C and 1D) and a dual pair of convex sets (Theorem 2), a mystical cancellation which is always present in the dual pair of spaces or operators (see formula (2.2)). In fact, we have kept the difficult part of the proofs out of this note. The main points of the proofs have already been published (references will be given at the appropriate places), however, then we missed some of the most interesting consequences and reformulations, and concentrated more on the technical value of the statements.

This main technical statement is the so-called Inverse Brunn-Minkowski inequality of the author [M1] (see also [M2], [P1], [P2]). Recall, that the Brunn-Minkowski inequality states that for any convex bodies K and T in R^n

$$\left[\operatorname{Vol}(K + T) \right]^{1/n} \geq (\operatorname{Vol} K)^{1/n} + (\operatorname{Vol} T)^{1/n} .$$

It was shown in [M1] that for every K there is an affine map $v_K \in \operatorname{SL}_n$ (v_K depends on K) and an affine image $\widehat{K} = v_K K$ of K (called a "main position" of K) such that considering

---

* Partially supported by a GIF grant

both bodies K and T in their suitable positions $\widehat{K}$ and $\widehat{T} = v_T T$ we will have

$$\left[\operatorname{Vol}(\widehat{K} + \widehat{T})\right]^{1/n} \le C\left[(\operatorname{Vol}K)^{1/n} + (\operatorname{Vol}T)^{1/n}\right] \tag{0.1}$$

where C is a universal constant (*independent*, of course, of n, K or T).

We see in this example that it is natural to associate to a convex body K the family $\mathcal{K}$ of the affine images $\{vK\}_{v \in \mathrm{GL}_n} = \mathcal{K}$. Every element $\widehat{K} = vK$ of this family induces a norm $\|\cdot\|_{\widehat{K}}$ of isometrically the same normed space $X = (\mathbb{R}^n, \|\cdot\|)$ and represents a different realization of space X. Therefore, it is natural, from a Functional Analysis point of view, to consider K together with the family $\mathcal{K}$. We call any such element $\widehat{K}$ a *position* of body K. However, it is probably surprising that, also from the pure geometric point of view, when a problem seems to be about an individual convex body, it is *natural* to consider this body K as an element of the family of its affine images $\mathcal{K}$ (the family of its *positions*). The inequality (0.1) was just the first step in realizing it but we meet this point of view more and more often, and we will see it throughout this paper. There are a few special positions of a convex body which play a very important (we could say, decisive) role in the whole asymptotic (by dimension going to infinity) theory of convexity.

A lot of unexpected results are direct and easy consequences of the inequality (0.1). Let us put some of them into this introduction. They are already quite well known to the experts in the field and, in some form, reflected in [M2] and Pisier's presentation [P1] of a proof of (0.1). But, in my experience, they did not find a way out of this narrow group, specially in the form in which we put them below. At the same time, they are definitely attractive for a much larger group of mathematicians.

Let K and T be arbitrary convex bodies in $\mathbb{R}^n$ and let $\operatorname{vol}K = \operatorname{Vol}T$. What can be said about a low bound for $\operatorname{Vol}(K \cap T)$? How small can it be? Similarly, how large can $\operatorname{Vol}\left[\operatorname{Conv}(K \cup T)\right]$ be? Of course, we can easily choose bodies K and T in such positions that their intersection will have a volume smaller than any given bound. However, if we want to estimate how different from the volume point of view two different bodies are, then we have to choose a special position for them which is good for such a comparison. And then, as we will see below, surprisingly there are no convex symmetric bodies in $\mathbb{R}^n$ for large n which are too different.

Proposition 0.2. *There are absolute constants $c > 0$ and C such that for any integer n and any $K \subset \mathbb{R}^n$ and $T \subset \mathbb{R}^n$ (convex symmetric bodies), $\operatorname{Vol}K = \operatorname{Vol}T$, there are positions*

$\widehat{K} = v_K K$ and $\widehat{T} = v_T T$, where v_K and $v_T \in \mathrm{SL}_n$ such that

$$\mathrm{Vol}(\widehat{K} \cap \widehat{T}) \geq c^n \, \mathrm{Vol}\, K$$

and

$$\mathrm{Vol}\left[\, \mathrm{Conv}(\widehat{K} \cup \widehat{T})\right] \leq C^n \, \mathrm{Vol}\, K \ .$$

So, any two convex symmetric bodies of the same volume cannot cut too much volume from each other in their respectively chosen *positions*.

Let us develop these facts to demonstrate better the role of a family of positions. As usual in this paper, K, T and P below are convex, centrally symmetric, compact bodies in $\mathbb{R}^n$; $K + T = \{x + y \mid x \in K \ , \ y \in T\}$ is the Minkowski sum of bodies K and T.

Theorem 0.3. *There is a universal constant C such that for every $K \subset \mathbb{R}^n$ some ellipsoid $\mathcal{M}_K$ corresponds to K, $\mathrm{Vol}\, K = \mathrm{Vol}\, \mathcal{M}_K$, and for every T*

$$\frac{1}{C^n} \, \mathrm{Vol}(\mathcal{M}_K + T) \leq \mathrm{Vol}(K + T) \leq C^n \, \mathrm{Vol}(\mathcal{M}_K + T) \ ,$$

$$\frac{1}{C^n} \, \mathrm{Vol}\left[\, \mathrm{Conv}(\mathcal{M}_K \cup T)\right] \leq \mathrm{Vol}\left[\, \mathrm{Conv}(K \cup T)\right] \leq C^n \, \mathrm{Vol}\left[\, \mathrm{Conv}(\mathcal{M}_K \cup T)\right] \ ,$$

$$\frac{1}{C^n} \, \mathrm{Vol}(\mathcal{M}_K \cap T) \leq \mathrm{Vol}(K \cap T) \leq C^n \, \mathrm{Vol}(\mathcal{M}_K \cap T) \ .$$

So, for every $K \subset \mathbb{R}^n$, a position $\mathcal{M}_K = u_K D$, $u_K \in \mathrm{GL}_n$, of the euclidean ball D is associated with K such that $\mathcal{M}_K$ can essentially substitute K in *any* volume computations. Moreover, the euclidean ball D does not play a special role here. *Any body $P \subset \mathbb{R}^n$* (as above) and the family of its position $\mathcal{P} = \{vP\}_{v \in \mathrm{GL}_n}$ can play the role of ellipsoids.

Remark 0.4. For every K there is a position $P_K = u_K P$, $u_K \in \mathrm{GL}_n$, $\mathrm{Vol}\, K = \mathrm{Vol}\, P_K$, such that P_K can be put in all inequalities of Theorem 0.3 instead of $\mathcal{M}_K$.

So, we have some (high dimensional) **Principle:**

The affine family of any fixed convex symmetric body (say, P) is rich enough to substitute any other body (K above) in an essential part of volume computations.

This reflects a new intuition, which we slowly learn about high dimension: instead of the expected increasing of essentially different bodies in $\mathbb{R}^n$ with increasing $n \to \infty$ (diversity of possibilities) we observe their *decreasing* to essentially one (any) body with its affine class.

Proofs of all the above mentioned results can either be found in [M2] or in the book [P1], or easily follow from the results stated and proved there.

In the first section of this paper we list some of results which use duality and give preliminary discussions which prepare us for the proofs and introduce the necessary terminology.

We describes some recent results in the second section and use them to prove the remaining statements of the previous section. We also add some information on the matter of duality.

Then, in the third section, we will apply our knowledge of duality to a study of the growth (in different directions) of a polynomial of many variables and to an estimation of a number of integer points in convex regions.

1. Geometric (Global) Duality Relations

Theorem 1.A. *There is a numerical constant C such that for any finite dimensional (say, n-dimensional) normed space $X = (\mathbb{R}^n, \|\cdot\|)$, there are two linear operators $u_1, u_2 \in SL_n$ such that if the norm $\||x\|| = \|x\| + \|u_1 x\|$ then the space Y with the norm $\||x\||^* + \||u_2 x\||^* = \|x\|_Y$ is euclidean up to the above (universal) constant C.*

This means, in a precise form, that the *Banach-Mazur distance*

$$d(Y, \ell_2^n) = \inf \left\{ \|T\| \cdot \|T^{-1}\| \mid T : Y \to \ell_2^n \text{ is a linear isomorphism} \right\} \leq C \ .$$

Note that we may realize X in the euclidean space $(\mathbb{R}^n, |\cdot|)$ (i.e. to consider a norm $\|\cdot\|_1$ on $\mathbb{R}^n$ such that X is isometric to $(\mathbb{R}^n, \|\cdot\|_1)$) in such a way that the above operators $u_i \in SO_n$ when $\|\cdot\|_1$ substitute $\|\cdot\|$.

Moreover, let $\mu^2 = \mu \times \mu$ be the Haar probability measure on $SO_n \times SO_n$ where μ is the Haar probability measure on SO_n; there is a set $\mathfrak{A} = \mathfrak{A}(X) \subset SO_n \times SO_n$ of pairs of orthogonal operators of "almost full" measure, i.e. $\mu^2(\mathfrak{A}) \geq a_n \to 1$, $n \to \infty$(and a sequence $\{a_n\}$ is independent of a choice of a sequence of normed spaces X), such that the statement of Theorem 1 is satisfied for any pair (u_1, u_2) of orthogonal operators from $\mathfrak{A}$. It is acceptable jargon to call such operators $u_i \in \mathfrak{A}$ "random" orthogonal operators meaning, of course, some "asymptotic randomness" when dimension increases to infinity.

We will also list a few reformulations of Theorem 1.A which use geometric language or a mixture of functional analysis and geometric languages.

Let K_1 and K_2 be convex centrally symmetric bodies in $\mathbb{R}^n$ which define the norms $\|\cdot\|_{K_1}$ and $\|\cdot\|_{K_2}$ being the unit balls of these norms.

Note that if T is the unit ball of the following norm $\|\cdot\|_T$

$$\|x\|_T = \|x\|_{K_1} + \|x\|_{K_2}$$

then the polar body T°, which is the unit ball of the dual norm $\left(\|\cdot\|_T\right)^*$, is the Minkowski sum of K_1° and K_2°

$$T^\circ = K_1^\circ + K_2^\circ \; .$$

So, the sum of convex bodies is the dual operation to the sum of norms. Consequently, Theorem 1.A can be reformulated in the following form.

Theorem 1.B. *Let X be an n-dimensional normed space with the unit ball $K \subset \mathbb{R}^n$. There are two linear operators $u_1, u_2 \in SL_n$ such that:*
define a convex body

$$T = K + u_1 K$$

(so T is the Minkowski sum of bodies K and $u_1 K$) and consider the norm $\|\cdot\|_T$; then the space Y with the norm $\|x\|_Y = \|x\|_T + \|u_2 x\|_T$ is euclidean up to the universal constant C mentioned in Theorem 1.A.

Again, as in Theorem 1.A, we may realize X in the euclidean space $(\mathbb{R}^n, |\cdot|)$, i.e. find some affine image $\widehat{K} = vK$, $v \in SL_n$, such that the above operators u_i could be chosen as "random" orthogonal operators.

Theorem 1.C. *Let $K \subset \mathbb{R}^n$ be a convex centrally symmetric, compact body (i.e. 0 is an inner point of K). There are two linear maps u_1 and $u_2 \in SL_n$ such that:*
define

$$T = \mathrm{conv}(K \cup u_1 K)$$

and consider

$$L = T \cap u_2 T \; ;$$

then L is ellipsoid $\mathcal{E}$ up to a universal (independent of n or $K \subset \mathbb{R}^n$) constant C. This means that the geometric (multiplicative) distance

$$d(L, \mathcal{E}) \overset{\text{def}}{=} \inf\{a \cdot b \mid L \subset a\mathcal{E} \quad \text{and} \quad \mathcal{E} \subset bL\} \leq C \; .$$

Again, there is a position $\widehat{K}$ of K, i.e. $\widehat{K} = vK$ for some special $v \in SL_n$, such that u_1 and u_2 may be taken as orthogonal rotations, and even "random" rotations. All these, of course, under assumption that K is substituted by $\widehat{K}$.

It is again acceptable jargon for the asymptotic theory to call such a body L, as appears in Theorem 1.C, an *isomorphic ellipsoid*. Trying to be exact in this jargon, we will say that, given number C, a family of convex bodies $\{L_n \subset \mathbb{R}^n \ , \ n \to \infty\}$ such that for some family of ellipsoids $\{\mathcal{E}_n \subset \mathbb{R}^n\}$, $d(L_n, \mathcal{E}_n) \leq C$, is a family of isomorphic ellipsoids. Similarly, we say "an isomorphic euclidean ball" meaning that all $\mathcal{E}_n = D_n = \{x \in \mathbb{R}^n \mid |x| \leq 1\}$ are the standard euclidean balls. By the way, we usually write D instead of D_n. In the last reformulation, we will use the polar operation explicitly.

Theorem 1.D. *For any $K \subset \mathbb{R}^n$ as above, there is a position $\widehat{K} = vK$, $v \in \mathrm{SL}_n$, and orthogonal rotations $u_1, u_2 \in \mathrm{SO}_n$ such that defining*

$$T = \widehat{K} + u_1 \widehat{K} \quad \text{and} \quad L = T^\circ + u_2 T^\circ \,.$$

we receive an isomorphic euclidean ball L, i.e. $d(L, D) \leq C$ where $D = \{x \in \mathbb{R}^n \mid |x| \leq 1\}$ is the euclidean ball and C is some universal constant, as in Theorems 1.A-C.

It is obvious that Theorem 1.D is equivalent to Theorem 1.B; it is enough to apply duality one more time between the sum of norms and the norm defined by the Minkowski sum of convex bodies.

It is also immediate that Theorem 1.C is equivalent to Theorem 1.D. To see this, observe that (for centrally symmetric convex bodies K_i)

$$\tfrac{1}{2}(K_1 + K_2) \subset \mathrm{Conv}(K_1 \cup K_2) \subset K_1 + K_2$$

and the convex hull of two such bodies is the dual operation to the intersection of the dual bodies:

$$(\mathrm{Conv}\, K_1 \cup K_2)^\circ = K_1^\circ \cap K_2^\circ \,.$$

So, all four statements above are just different forms of the same theorem. Note that in all four forms, A-B-C-D, Theorem 1 shows that applying two dual operations to an arbitrary convex body, we may regularize it to be close to an ellipsoid. In fact, after this main step, we may approximate an ellipsoid up to any $\varepsilon > 0$ just adding a few more rotations where the number of rotations needed depends on $\varepsilon > 0$ but not on dimension n or body $K \subset \mathbb{R}^n$. This is already a consequence of another theorem (see [BLM], Theorem 6):

Let $K \subset \mathbb{R}^n$ be a convex (centrally symmetric) compact body and $d(K, D) \leq C$ where D is the unit euclidean ball in $\mathbb{R}^n$. Then for every $\varepsilon > 0$ there is $p(\varepsilon) \sim (C/\varepsilon)^2$ rotations $\{u_i \in \mathrm{SO}_n\}_{i=1}^{p(\varepsilon)}$ such that

$$T = \frac{1}{p(\varepsilon)} \sum_{i=1}^{p(\varepsilon)} u_i K$$

is "almost" the euclidean ball: $d(T, D) \leq 1 + \varepsilon$. Similarly, of course, the norm $\|x\|_T = \frac{1}{p(\varepsilon)} \sum_{i=1}^{p(\varepsilon)} \|u_i x\|_K$ is, up to $(1 + \varepsilon)$, euclidean.

We will prove Theorem 1 in the form of 1.B. We announced the theorem in [M3], in the form of 1.D. I would like now to discuss some associations of this theorem, including an historical one.

In 1976, Kašin [K] proved that for any integer n there is an orthogonal rotation $u \in SO_n$ such that for any $x \in \mathbb{R}^n$

$$\frac{1}{C} \|x\|_{\ell_2} \leq \|x\|_{\ell_1} + \|ux\|_{\ell_1} \leq C \|x\|_{\ell_2}$$

where C is a universal constant (independent of n!) and, as usual, for $x = (x_i)_1^n \in \mathbb{R}^n$, $\|x\|_{\ell_1} = \sum |x_i|$ and $\|x\|_{\ell_2} = \sqrt{\sum |x_i|^2}$. Let $C^n = [-1, 1]^n$ be the n-dimensional cube. Then the dual statement to Kašin's theorem says that there is an orthogonal operator $u \in SO_n$ such that

$$T = C^n + uC^n$$

is uniformly (by n) close to the euclidean ball: $d(T, D) \leq C$ for some universal constant C. In fact, he also proved that such operators $u \in SO_n$ form an asymptotically, when $n \to \infty$, almost full Haar (probability) measure on SO_n.

So, in the case of K being the cube C^n, only one step is needed in Theorem 1.D to derive an isomorphic euclidean ball; we don't need the second step and a construction of a body L.

Could this be the case for any K? Of course not. If we take K as the unit ball of space ℓ_1, i.e. $K = \text{Conv}\{\pm e_i\}_1^n$, $\{e_i\}_1^n$ is the natural orthonormal basis of $\ell_2^n = (\mathbb{R}^n, |\cdot|)$, which also means that K is the polar of the cube C^n, then at least $N \sim n/\log n$ rotations $\{u_i\}_1^N$ are needed to receive a body $T = \frac{1}{N} \sum_1^N u_i K$ which has a uniformly bounded (by $n \to \infty$) distance from a euclidean ball (or any other ellipsoid). We investigated this question in [BLM]. It was also shown there that such a number of rotations ($N \sim n/\log n$) is already enough for *any* convex body $K \subset \mathbb{R}^n$: for any convex body $K \subset \mathbb{R}^n$ there is a position $\widehat{K} = vK$, $v \in SL_n$, and a set of orthogonal rotations $\{u_i\}_{i=1}^N \subset SO_n$ for $N \sim n/\log n$ such that

$$T = \frac{1}{N} \sum_1^N u_i \widehat{K}$$

is an isomorphic ellipsoid, i.e. for some universal constant C (independent of n or $K \subset \mathbb{R}^n$) $d(T, D) \leq C$.

To investigate a level of regularization which we can achieve by only a few (say, two) rotations, we introduce the numbers

$$r_\ell(K) = \max\left\{ r \mid rD \subset \frac{1}{\ell} \sum_1^\ell u_i K \ , \ u_i \in \mathrm{SO}_n \right\}.$$

We prove the following theorem in section 2.

Theorem 2. *There is a number $c > 0$ such that for any integer n, any convex, centrally symmetric, compact body $K \subset \mathbf{R}^n$*

$$c \leq r_2(K) \cdot r_3(K^\circ)$$

where K° is the polar body of K. (Moreover, we show that $c > \frac{1}{21}\sqrt{\frac{2}{3}} \simeq 0.03888$.)

We do not assume any special position for K in this theorem and therefore say that in any position and, for any K, either K itself or its dual K° can be well regularized after one or two rotations.

Note that the Kašin theorem is an immediate consequence of Theorem 2: let $K = C^n$; it is easy to see that $r_3(K^\circ) \leq \frac{C}{\sqrt{n}}$ for some constant C (just by estimating from above the volume of the convex hull $\mathrm{Conv}\{\pm x_i\}_{i=1}^{3n}$ of any $3n$ points $\{x_i\} \subset S^{n-1} = \partial D$); then $r_2(C^n) \geq c_1\sqrt{n}$ which means that for some $u \in \mathrm{SO}_n$

$$c_1\sqrt{n}D \subseteq \frac{C^n + uC^n}{2} = T.$$

Also, obviously, $T \subset \sqrt{n}D$ and therefore $d(T, D) \leq 1/c_1$.

We also extend Theorem 2 in section 2 to involve the duality property of $r_\ell(K)$ for any integer ℓ. Theorem 3 is proved there.

Theorem 3. *Let K be a convex centrally symmetric compact body. Fix any $\kappa > 0$ and let $\lambda = 1 - \frac{1}{\ell} - \kappa$ for an integer $\ell > 1$. Then there is a $[\lambda n]$-dimensional subspace E such that*

$$K \cap E \subseteq r_{\ell_1}(K)\frac{\sqrt{\ell}}{\kappa}D$$

where ℓ_1, $\ell \leq \ell_1 < 2\ell$, is a power of two.

We will also discuss a few possibilities of strengthening this theorem and reformulate it in the language of approximation numbers for any operator $A : X \to \ell_2^n$.

In our plan, we derive Theorem 1.B from Theorem 2 above. The main step here is the use of the so-called Inverse Brunn-Minkowski inequality of the author [M1] (see also [M2], [P1]).

We recalled in the Introduction, the Brunn-Minkowski inequality and what we mean by its inverse – inequality (0.1). We saw there that to satisfy the inverse inequality we have to put every body in a special position. Different properties of this position (sometimes called the main position or m-position) were also discussed in the papers cited above; we will need two of them:

one which says that if $\widehat{K}$ is already in our main position and $u \in O_n$ then $u\widehat{K}$ is also in its main position,

and another property which claims that $(\widehat{K})^\circ$ is also in its main position (under the assumption that $\widehat{K}$ was put in the main position).

Now return to Theorem 2 and apply it to $\widehat{K} = vK$ – the main position of K. Then, for some $u \in SO_n$,

$$r_2(\widehat{K})^n \operatorname{Vol} D \leq \operatorname{Vol}\left(\frac{\widehat{K} + u\widehat{K}}{2}\right) \leq C^n \operatorname{Vol} \widehat{K}$$

and similarly

$$r_3(\widehat{K}^\circ)^n \operatorname{Vol} D \leq C_1^n \operatorname{Vol} \widehat{K}^\circ$$

(C and C_1 are numerical constants).

Then, from one side, Theorem 2 implies that for some universal number $c_1 > 0$

$$c_1^n \leq \frac{\operatorname{Vol} \widehat{K} \cdot \operatorname{Vol} \widehat{K}^\circ}{(\operatorname{Vol} D)^2} = \frac{\operatorname{Vol} K \cdot \operatorname{Vol} K^\circ}{(\operatorname{Vol} D)^2},$$

which is the so-called Inverse Santaló Inequality, proven by Bourgain and the author [BM], and, from the other side, by the Santaló inequality

$$r_2(\widehat{K}) \cdot r_3(\widehat{K}^\circ) \leq C \cdot C_1 \left[\frac{\operatorname{Vol} \widehat{K} \cdot \operatorname{Vol} \widehat{K}^\circ}{(\operatorname{Vol} D)^2}\right]^{1/n} \leq C \cdot C_1 = C_2.$$

This shows that there is a number C_3 (again, of course, independent of n or $K \subset \mathbb{R}^n$) such that

$$\operatorname{Vol} \frac{\widehat{K} + u\widehat{K}}{2} \leq C^n \operatorname{Vol} \widehat{K} \leq C_3^n r_2(\widehat{K})^n \operatorname{Vol} D$$

(and, by the same reasoning,

$$\operatorname{Vol} \widehat{K}^\circ \leq C_3^n r_2(\widehat{K}^\circ)^n \operatorname{Vol} D).$$

Define $T = \frac{\widehat{K} + u\widehat{K}}{2}$. Then $r_2(\widehat{K})D \subset T$ and $\operatorname{Vol} T \leq C_3^n \operatorname{Vol}\left(r_2(\widehat{K})D\right)$. We summarize this fact in the following statement. We use the following standard terminology: Let $\mathcal{E} \subset T$ be the maximal volume ellipsoid inscribed in a convex body T; we call the volume ratio

$$\text{v.r.}\, T = (\operatorname{Vol} T / \operatorname{Vol} \mathcal{E})^{1/n}.$$

Theorem 4. *There is a universal constant C such that for any convex centrally symmetric convex body $K \subset \mathbb{R}^n$ there is an operator $u \in \mathrm{SL}_n$ such that a convex body T*

$$T = K + uK$$

has the volume ratio at most C.

Remark. In the proof of Theorem 4 from Theorem 2, we considered first the position $\hat{K}$ of K then used an orthogonal rotation $u \in \mathrm{SO}_n$. Because of the probability nature of the proof of Theorem 2, we prove, in fact that there is a "large amount" of $u \in \mathrm{SO}_n$ such that $T = \hat{K} + u\hat{K}$ has the volume ratio bounded by C. In particular, it enables proving that the same $u \in \mathrm{SO}_n$ can be chosen for $\hat{K}$ and $\hat{K}°$.

In section 2 we will show that Theorem 4 implies Theorem 1.B.

2. Linear Duality Relations and Proofs of the Theorems

Consider a normed space $X = (\mathbb{R}^n, \|\cdot\|)$ and $|\cdot|$ denote the standard euclidean norm in $\mathbb{R}^n$. Let, for any $x \in \mathbb{R}^n$,

$$\frac{1}{a}|x| \le \|x\| \le b|x| .$$

Then the dual space X^* has the norm $\|x\|^* = \sup\{|(x,y)| \mid \|y\| \le 1\}$. We denote, as usual, $K = \{x \in \mathbb{R}^n \mid \|x\| \le 1\}$ being the unit ball of X, $D = \{x \in \mathbb{R}^n \mid |x| \le 1\}$ – the euclidean unit ball and $S^{n-1} = \partial D = \{x \in \mathbb{R}^n \mid |x| = 1\}$. We always equip S^{n-1} with the rotation invariant probability measure μ. The following two integral characteristics of a relative position of a norm $\|\cdot\|$ and a euclidean norm $|\cdot|$ play the central role in our theory

$$M = \int_{x \in S^{n-1}} \|x\| d\mu(x) \quad \text{and} \quad M^* = \int_{x \in S^{n-1}} \|x\|^* d\mu(x) .$$

To emphasize the role of space X, we may write $M(X)$ or M_K instead of just M. Then, for example, $M(X^*) = M^*$ and is the same as $M_{K°}$ where $K°$ is the polar of K and is the unit ball of the dual space X^*.

Introduce also the following integer function on $\mathbb{R}^+$ associated with $X = (\mathbb{R}^n, \|\cdot\|, |\cdot|)$:

$$t(r) = t(X; r) = \sup\{k \in \mathbb{N} \mid \exists \text{ a subspace } E \hookrightarrow \mathbb{R}^n \text{ of dimension } k \text{ such that}$$
$$|x| \le r\|x\| \text{ for every } x \in E\} .$$

Similarly we write $t^*(r) = t(X^*; r)$.

We proved in [M4] that for any space X, for any $\kappa > 0$ and any $r > 0$

$$t(r) + t^* \left(\frac{1}{\kappa r}\right) \geq (1 - \kappa)n - C , \tag{2.1}$$

for some universal constant $C > 0$ (independent of anything). This fact reflects some deep linear duality relations and will be the basic tool in proving Theorem 2.

In fact, we will need a probability version of inequality (2.1). We achieve this by changing the definition of $t(r)$: we will now ask for the existence of a set of k-dimensional subspaces of a large measure instead of, probably, the only subspace. To put it in an exact form, we have to decide what we are going to mean by "a large measure". There are different possibilities and I choose only what will be used later on.

Let $G_{n,k}$ be the Grassmannian manifold of all k-dimensional subspaces of $\mathbb{R}^n$; it is a homogeneous space of O_n and we equip it with the Haar probability measure ν. Define

$\bar{t}(r) = \bar{t}(X; r) = \sup \{k \in \mathbb{N} \mid \exists \mathfrak{A} \subset G_{n,k} ,$ of a probability $\nu(\mathfrak{A}) > 1 - \frac{1}{n}$ such that for every subspace $E \in \mathfrak{A}, |x| \leq r\|x\|$ for every $x \in E\}$.

Similarly, we define $\bar{t}^*(r)$. Then (see [M4]), for any space X, any $\kappa > 0$ and any $r > 0$

$$\bar{t}(r) + \bar{t}^* \left(\frac{1}{\kappa r}\right) \geq (1 - \kappa)n - o(n) \tag{2.1'}$$

for some (universal) function $o(n)$ (i.e. $\frac{o(n)}{n} \to 0$ when $n \to \infty$).

Remark – Correction. The paper [M4] contains a mistake in normalization in the outline of a probability variant (2.1') of inequality (2.1). I refer to Proposition 2.ii and the preceding Remark in [M4]. I will correct this mistake here. In fact, that Remark and the following Proposition 2.ii, are just an adjustment of Corollary 3.4 from Gordon [G] to our needs. I put them now in the corrected forms:

Define $a_k = \sqrt{2}\Gamma\left(\frac{k+1}{2}\right)/\Gamma\left(\frac{k}{2}\right)$. Let $0 < \theta < 1, 0 < \lambda < 1$ and $k = [\lambda n]$. There is a set $A \subset G_{n,k}$ of k-dimensional subspaces such that for every $E \in A$

$$\frac{\theta a_{n-k}}{a_n M^*}|x| \leq \|x\| \quad \text{for every} \quad x \in E$$

and $\mu(A) \geq 1 - \frac{7}{2}\exp\left(-\frac{1}{18}a_{n-k}^2(1-\theta)^2\right)$.

Choosing $\theta = 1 - \frac{1}{\sqrt{k}}$ and, the second time, $\theta = 1 - \frac{1}{k^{1/3}}$, we receive from the above estimate the corrected form of

Proposition 2.ii from [M4].

1. *For any* $X = (\mathbb{R}^n, \|\cdot\|, |\cdot|)$ *any* λ, $0 < \lambda < 1$, *and*

$$k = \left[\lambda n - 2(1-\lambda)\frac{n}{\sqrt{k}} + O\left(\frac{n}{k}\right)\right] = \begin{cases} \lambda n - O\left(\sqrt{\frac{n}{\lambda}} + \frac{1}{\lambda}\right) & \text{for } \lambda \text{ small} \\ \lambda n - O(\sqrt{n}) & \text{for } \lambda \geq \varepsilon > 0, \end{cases}$$

there is a set $\mathcal{A} \subset G_{n,k}$ *such that for every* $E \in \mathcal{A}$

$$\frac{\sqrt{1-\lambda}}{M^*}|x| \leq \|x\| \qquad \text{(for every } x \in E\text{)}, \qquad (*)$$

and $\mu(\mathcal{A}) \geq 1 - c_1 \exp\left(-c_2 \frac{n-k}{k}\right)$, *where* μ *is the Haar probability measure on* $G_{n,k}$ *and* c_1, c_2 *are some positive universal constants.*

2. *Similarly, for* $k = \left[\lambda n - 2(1-\lambda)\frac{n}{k^{1/3}} - 4(1-\lambda)\frac{n}{k^{2/3}} + O(1)\right]$ *there is a set* $\mathcal{A} \subset G_{n,k}$ *such that every* $E \in \mathcal{A}$ *is satisfied* $(*)$ *and* $\mu(\mathcal{A}) \geq 1 - c_1 \exp\left(-c_2 \frac{n-k}{k^{2/3}}\right)$ *where constants* c_1 *and* c_2 *are universal numbers.*

With these corrections, the proof of (2.1') can be read in [M4].

We now return to the discussion of inequality (2.1).

For experts more equipped with the language of approximation numbers I will rewrite (2.1) in the terms of Gelfand numbers of operator $A : X \to \ell_2^n$. Then

$$c_k(A) = \inf\left\{\|A|_E\| \mid E \text{ is a subspace of } \mathbb{R}^n, \text{ codim } E = k\right\},$$

where $\|A|_E\|$ is, of course, the norm of operator A restricted on subspace E, is called k-th Gelfand number of A. Clearly, if A is the identity map $(\mathbb{R}^n, \|\cdot\|) = X \to (\mathbb{R}^n, |\cdot|)$ then

$$t(c_k(A)) = n - k.$$

Also, for any invertible $A : X \to \ell_2^n$, $(A^{-1})^* : X^* \to \ell_2^n$ and (2.1) means that for any $\kappa > 0$ and any integer k, $1 \leq k \leq n$,

$$c_k(A) \cdot c_{n(1+\kappa)-k+C}(A^{*-1}) \leq \frac{1}{\kappa}, \qquad (2.2)$$

where C is a universal constant. I would like to emphasize that the strength of this inequality in the fact that it is valued for *all* the range of k from 1 to n. At the end of this section we show another type of estimate on $c_k(A)$. The proof of inequality (2.1) uses, besides other tools, the following

M^*-low bound:

For any $X = (\mathbb{R}^n, \|\cdot\|, |\cdot|)$ and any λ, $0 < \lambda < 1$, there is a $k = [\lambda n]$-dimensional subspace E_k such that for any $x \in E_k$

$$\frac{f(\lambda)}{M^*}|x| \leq \|x\| , \tag{2.3}$$

It was originally shown in [M5] that one may take $f(\lambda) \geq 1 - \lambda + o(\lambda)$ when $\lambda \to 0$. It was improved by Pajor and Tomčzak [PT] in the case of $\lambda \to 1$ to $f(\lambda) \geq c\sqrt{1-\lambda}$ for some $c > 0$ and was further improved by Gordon [G] to $f(\lambda) \geq \sqrt{1-\lambda}\left(1 + O\left(\frac{1}{(1-\lambda)n}\right)\right)$ (i.e. $c \simeq 1$). The last form of estimate on $f(\lambda)$ is used to prove (2.1). Estimate (2.3) also gives, of course, an estimate on Gelfand's number of an operator $A : X \to \ell_2^n$ which is in the best form of Pajor-Tomčzak

$$c_k(A) \lesssim \sqrt{\frac{n}{k}} M^*$$

where M^* here is $M^* = \int\limits_{x \in S^{n-1}} \|A^*x\| d\mu(x)$. This individual estimate on Gelfand's numbers of operator A alone is, in fact, asymptotically the best possible. However, joining it with the similar estimate for $c_{k_1}(A^{*-1})$, we receive

$$c_k(A) \cdot c_{k_1}(A^{*-1}) \leq \frac{n}{\sqrt{k \cdot k_1}} MM^*$$

which is much worse than (2.2) in almost all possible cases. So, the strength of (2.2) (and, which is the same, (2.1)) in mystical cancellations which are always present in the dual pair of spaces, bodies or operators.

Proof of Theorem 2. Consider r such that $\bar{t}(r) = \frac{n}{2}$ (let n be even). Then there are two subspaces E_1 and E_2, $\dim E_i = n/2$, orthogonal in $(\mathbb{R}, |\cdot|)$, i.e. $E_1 \oplus E_2 = \mathbb{R}^n$, and such that

$$\frac{1}{r}|x| \leq \|x\| \quad \text{for} \quad x \in E_1 \cup E_2 . \tag{2.4}$$

I borrow the next reasoning from Krivine's proof of the above mentioned Kašin theorem (as explained to me by Pisier – see [P1]). Let P_i be the orthogonal projection $\mathbb{R}^n$ onto E_i. Then the identity map $I = P_1 + P_2$. Consider an orthogonal map $T = P_1 - P_2$. Then, for any $x \in \mathbb{R}^n$, $x = x_1 + x_2$ where $x_i = P_i x \in E_i$ and

$$\|x_1 + x_2\| + \|x_1 - x_2\| \geq 2 \max_{i=1,2} \|x_i\| \geq \text{(by (2.4))}$$

$$\geq \frac{2}{r} \max_{i=1,2} |x_i| \geq \frac{2}{\sqrt{2}r}\sqrt{|x_1|^2 + |x_2|^2} = \frac{2|x|}{r\sqrt{2}} .$$

Therefore,

$$\frac{\|x\| + \|Tx\|}{2} \geq \frac{1}{\sqrt{2}r}|x| .$$

By duality, this means

$$\frac{K^\circ + TK^\circ}{2} \supset \frac{1}{\sqrt{2}r}D \ .$$

So, if $\bar{t}(X,r) \geq n/2$ then

$$r_2(K^\circ) \equiv r_2^* \geq \frac{1}{\sqrt{2}r} \ .$$

(similarly, of course, $\bar{t}(X^*,\rho) \geq \frac{n}{2}$ implies $r_2 \geq \frac{1}{\sqrt{2}\rho}$). Unfortunately, we cannot achieve both $\bar{t}(r)$ and $\bar{t}^*\left(\frac{1}{\kappa r}\right)$ being $\geq \frac{n}{2}$. But, choosing $\kappa =$, say, $\frac{1}{7} < \frac{1}{6}$, we have $\bar{t}^*\left(\frac{7}{r}\right) \geq n/3$ (starting with n large enough; we do not repeat this condition later on). Then, again by a probability argument, we find three subspaces $E_i \ i = 1,2,3$, pairwise orthogonal, $E_1 \oplus E_2 \oplus E_3 = \mathbb{R}^n$, $\dim E_i = n/3$ (or, if one does not like to assume that $n/3$ is an integer, one may take $\dim E_i =$ either $[n/3]$ or $[n/3]+1$) and

$$\kappa \cdot r \cdot |x| \leq \|x\|^* \quad \text{for} \quad x \in \bigcup_1^3 E_i \ ; \tag{2.5}$$

(we choose $\kappa = 1/7$ but will continue to write κ). Let P_i be the orthogonal projection onto E_i and write any $x \in \mathbb{R}^n$

$$x = x_1 + x_2 + x_3$$

where $x_i = P_i x \in E_i$. Consider orthogonal operators $T_1 = P_1 - P_2 - P_3$ and $T_2 = P_1 + P_2 - P_3$. Then, by the triangle inequality, it is easily seen that

$$2\|x_1\|^* \leq \|x\|^* + \|T_1 x\|^*$$

and

$$2\|x_2 + x_3\|^* \leq \|x\|^* + \|T_1 x\|^* \ , \ 2\|x_2 - x_2\|^* \leq \|T_1 x\|^* + \|T_2 x\|^*$$

which implies for $i = 2,3$

$$4\|x_i\|^* \leq \|x\|^* + 2\|T_1 x\|^* + \|T_2 x\|^* \ .$$

To put it in a uniform way, we weaken these inequalities to

$$\frac{2}{3}\|x_i\|^* \leq \frac{1}{3}\left(\|x\|^* + \|T_1 x\|^* + \|T_2 x\|^*\right) \quad \text{for} \quad i = 1,2,3 \ .$$

Using (2.5), we obtain

$$\frac{2}{3}\kappa \cdot r\frac{|x|}{\sqrt{3}} \leq \frac{\|x\|^* + \|T_1 x\|^*\| + \|T_2 x\|^*}{3} \ .$$

Dualizing, we receive

$$\frac{K + T_1 K + T_2 K}{3} \geq \frac{2}{3\sqrt{3}} \kappa r \, D \, .$$

Therefore, if $\bar{\ell}\left(X^*, \frac{1}{\kappa r}\right) \geq \frac{n}{3}$ then

$$r_3(K) = r_3 \geq \frac{2}{3\sqrt{3}} \kappa r \, .$$

Finally, we have

$$r_2(K^\circ) \cdot r_3(K) \geq \frac{1}{3}\sqrt{\frac{2}{3}}\kappa = \frac{1}{21}\sqrt{\frac{2}{3}} \, . \qquad \qquad \square$$

We will extend the previous reasoning to the case where we have a given bound r between a norm and the euclidean norm on a "random" n/ℓ-dimensional subspace. So, we consider now r such that $\bar{\ell}(r) =]n/\ell[$, i.e. the closest integer to n/ℓ which is not smaller than n/ℓ. Then, by the standard probability argument, we find ℓ subspaces E_i, $i = 1, \ldots, \ell$, pair-wise orthogonal, $\oplus \sum_1^\ell E_i = \mathbb{R}^n$ and $\dim E_i$ is equal to either $]n/\ell[$ or $[n/\ell]$. Also on every one of these subspaces

$$\frac{1}{r}|x| \leq \|x\| \qquad \text{for} \quad x \in \bigcup_{i=1}^\ell E_i \, . \qquad (2.4')$$

Note that our main interest remains with the case ℓ a fixed number and $n \to \infty$. Let us first settle the case of $\ell = 2^t$ is a power of 2. Then consider the Walsh matrices of signs $W_t = (w_{ji})_{j,i=1}^{2^t}$. These matrices are built by induction:

$$W_1 = \begin{pmatrix} 1 & 1 \\ 1 & -1 \end{pmatrix} \qquad \text{and} \qquad W_t = \begin{pmatrix} W_{t-1} & W_{t-1} \\ W_{t-1} & -W_{t-1} \end{pmatrix} \qquad \text{for every } t \geq 2 \, .$$

Obviously, $\frac{1}{\sqrt{\ell}} W_t \in O(\ell)$ and, for every $i = 1, \ldots, \ell$, there is a vector of signs $\bar{\varepsilon}^{(i)} = (\varepsilon_j^{(i)} = \pm 1)_{j=1}^\ell$ such that

$$\sum_{j=1}^\ell \varepsilon_j^{(i)} w_{jk} = \begin{cases} \ell & \text{for } k = i \\ 0 & \text{for } k \neq i \, . \end{cases} \qquad (2.6)$$

Let P_i now be the orthogonal projection onto E_i and define $P_i x = x_i$. Define orthogonal operators W_j, $j = 1, \ldots, \ell$,

$$W_j x = \sum_{i=1}^\ell w_{ji} P_i x \, .$$

(Then, e.g. $W_1 = I$ is the identity operator). By (2.6), $\sum_j \varepsilon_j^{(i)} W_j x = \ell x_i$ and therefore for any $i = 1, \ldots, \ell$

$$\frac{1}{\ell} \sum_{j=1}^\ell \|W_j x\| \geq \|x_i\| \geq (\text{by } (2.4')) \; \frac{1}{r}|x_i| \, .$$

As $|x| = \sqrt{\sum |x_i|^2}$ and $\max |x_i| \geq \frac{1}{\sqrt{\ell}}|x|$, we obtain for every $x \in R^n$

$$\frac{1}{\ell} \sum_{j=1}^{\ell} \|W_j x\| \geq \frac{1}{r} \cdot \frac{1}{\sqrt{\ell}} |x| \ .$$

By duality this implies

$$\frac{1}{\ell} \sum_{1}^{\ell} W_j K^\circ \supset \frac{1}{r\sqrt{\ell}} D \ .$$

Therefore, we arrive at the following (preliminary) fact.

Fact. Let $\bar{\iota}(X, r) =]n/\ell[$ for ℓ being a power of 2. Then

$$r_\ell(K^\circ) \geq \frac{1}{\sqrt{\ell r}}$$

where, as usual, K° is the polar body to the unit ball K of X, i.e. K° is the unit ball of the dual space X^*.

In the case of arbitrary ℓ (not necessarily a power of 2), take the largest power of 2, $\ell_0 \leq \ell$ and introduce (if $\ell > \ell_0$)

$$y_i = x_{2i-1} + x_{2i} \qquad \text{for} \quad 1 \leq i \leq \ell - \ell_0 \ ,$$

$$y_i = x_{i+\ell-\ell_0} \qquad \text{for} \quad \ell_0 \geq i > \ell - \ell_0 \ .$$

Then $\forall x = \sum_{i=1}^{\ell_0} y_i$ and

$$W_j x = \sum_{i=1}^{\ell_0} w_{ji} y_i \ .$$

Introduce another orthogonal operator A

$$Ay_i = x_{2i-1} - x_{2i} \quad \text{for} \quad i = 1, \ldots, \ell - \ell_0 \quad \text{and} \quad Ay_i = y_i \quad \text{for} \quad i > \ell - \ell_0 \ .$$

Then, as we have proved above

$$\frac{1}{\ell_0} \sum_{1}^{\ell_0} \|W_j x\| \geq \max_{1 \leq i \leq \ell_0} \|y_i\|$$

and, similarly,

$$\frac{1}{\ell_0} \sum_{1}^{\ell_0} \|W_j A x\| \geq \max_i \|Ay_i\| \ .$$

Also, for $1 \leq i \leq \ell - \ell_0$,

$$\|Ay_i\| + \|y_i\| \geq 2 \max \left\{ \|x_{2i-1}\|, \|x_{2i}\| \right\} \ ,$$

which implies

$$\frac{\sum_1^{\ell_0} \|W_j x\| + \sum_1^{\ell_0} \|W_j A x\|}{2\ell_0} \geq \max_{1 \leq i \leq \ell} \|x_i\| \geq$$

(by (2.4'))

$$\geq \frac{1}{r} \max |x_i| \geq \frac{1}{r\sqrt{\ell}} |x| .$$

By duality, this means that for some orthogonal operators $T_j \in O_n$, $j = 1, \ldots, 2\ell_0$,

$$\frac{1}{2\ell_0} \sum_{j=1}^{2\ell_0} T_j K^\circ \supset \frac{1}{r\sqrt{\ell}} D .$$

So, the following lemma is proved.

Lemma 2.1. *Let $\bar{r}(X, r) =]n/\ell[$ and let $\ell_1 \geq \ell$ be the smallest power of two which is at least ℓ. Let K be the unit ball of X and K° be its polar, i.e. K° is the unit ball of the dual space X^*. Then*

$$r_{\ell_1}(K^\circ) \geq \frac{1}{r\sqrt{\ell}} . \tag{2.7}$$

Remark. Unfortunately, we have some increase of ℓ_1 with respect to ℓ: $\ell \leq \ell_1 < 2\ell$. It is not always necessary. If, for example, ℓ is an Hadamard number then we could take $\ell_1 = \ell$. Recall, that ℓ is called a Hadamard number if there is an orthogonal matrix $A \in O_\ell$ with all entries being $\pm 1/\sqrt{\ell}$. Note that, with increasing ℓ, most numbers of the form $\ell = 4k$ become Hadamard numbers (see [WSW]). Also in the case of any (not necessary Hadamard) ℓ we can write (2.7) with $\ell_1 = \ell$ but up to some constant factor. We did this, for example, in the case $\ell = 3$ to prove Theorem 2.

Straightforward use of Lemma 2.1 and theorem-inequality (2.1) leads to the following theorem.

Theorem 3.a. *Let K be a convex centrally-symmetric compact body. Fix $\kappa > 0$. Let $\lambda = 1 - \frac{1}{\ell} - \kappa$ for some integer $\ell > 1$. Then there is a $[\lambda n]$-dimensional subspace E such that*

$$K \cap E \subseteq r_{\ell_1}(K) \frac{\sqrt{\ell}}{\kappa} D \tag{2.8}$$

where ℓ_1, $\ell \leq \ell_1 < 2\ell$, is a power of two.

Choose now a special $\kappa = 1/\ell$. In this case, we in fact may use an addition to inequality (2.1) – see Theorem 1 from [M4] – which says that under some conditions one can put $\sqrt{\kappa}$ in (2.8). Therefore

Theorem 3.b. *Let* $\lambda = 1 - \frac{2}{\ell}$ *for an integer* $\ell > 2$ *and* K *as in Theorem 3a. Then there is a* $[\lambda n]$*-dimensional subspace* $E \subset \mathbb{R}^n$ *such that*

$$K \cap E \subsetneq a(\ell)\ell \cdot r_{\ell_1}(K)D$$

where $a(\ell) \to 1$ *for* $\ell \to \infty$ *(and* $a(\ell)$ *is independent of* n *or* $K \subset \mathbb{R}^n$*) and* ℓ_1 *has the same meaning as in Theorem 3a.*

The last reformulation of Theorem 3 will involve Gelfand's numbers $c_k(A)$ of an operator $A : X \to \ell_2^n$ which we have defined at the beginning of this section. Let X be any normed space, $\dim X \geq n$ and $A : X \to \ell_2^n$ be a linear operator onto. Let $K(X) = \{x \in X \mid \|x\| \leq 1\}$. Consider $AK(X) = K \subset \mathbb{R}^n$. Define the following "approximation" numbers of operator A:

$$r_\ell(A) = \max\left\{r \;\middle|\; rD \subset \frac{1}{\ell}\sum_{i=1}^\ell u_i K , \quad u_i \in O_n\right\},$$

where, as usual, $D = \{x \in \ell_2^n \mid |x| \leq 1\}$ is the unit euclidean ball. Then

Theorem 3.c. *In the situation described above*

$$c_{\frac{2}{7}n}(A) \leq a(\ell) \cdot \ell \cdot r_{\ell_1}(A)$$

where $a(\ell) \to 1$ *(* $\ell \to \infty$ *) and* ℓ_1, $\ell_1 < 2\ell$, *as in Theorem 3.b.*

Remark. The remark which preceded Theorem 3 can be applied to all the different formulations of the theorem. This means that for ℓ being an Hadamard number, we can take $r_\ell(K)$ instead of $r_{\ell_1}(K)$. Also, for any ℓ, we could take $r_\ell(K) \cdot c(\ell)$ instead of $r_{\ell_1}(K)$.

Proof of Theorem 1.B. We will now derive Theorem 1.B from Theorem 4. Note that this is, by now, a completely standard argument of Asymptotic Theory.

We start with a convex centrally symmetric body $T \subset \mathbb{R}^n$ which has the volume ratio bounded by C (recall that $T = K + uK$ where K is the unit ball of a space X). We can also assume that the standard unit euclidean ball $D \subset \mathbb{R}^n$ is the maximal volume ellipsoid of T. Then $D \subset T$ and

$$\frac{\mathrm{Vol}\, T}{\mathrm{Vol}\, D} = \int\limits_{z \in S^{n-1}} \frac{1}{\|x\|_T^n} d\mu(x) \leq C^n . \tag{2.9}$$

Here $\|\cdot\|_T$ is defined by $T = \{x \in \mathbb{R}^n \mid \|x\|_T \leq 1\}$ and $\mu(x)$ is the probability rotation invariant measure on the unit euclidean sphere $S^{n-1} = \partial D$. It was shown by Szarek-Tomčzak [ST] as an extension of the aforementioned Kašin result [K] that *a "random"* $]\frac{n}{2}[$-*dimensional subspace of* $X_T = (\mathbb{R}^n, \|\cdot\|_T)$ *is* C_1-*euclidean,* i.e. it has the distance from $(]\frac{n}{2}[)$-dimensional euclidean space bounded by C_1 where C_1 depends on C only. In fact, they introduced a notion of volume ratio (and spaces with finite volume ratio) for this purpose. I outline the main point of this reasoning. Fix $t > C$. One has from (2.9):

$$C^n \geq \int_{S^{n-1}} \frac{1}{\|x\|_T^n} d\mu(x) \geq \int_{\substack{x \in S^{n-1} \\ \|x\|_T \leq 1/t}} \frac{1}{\|x\|_T^n} d\mu(x) \geq t^n \mu\{x \in S^{n-1} \mid \|x\|_T \leq 1/t\} . \qquad (2.10)$$

Therefore

$$\mu\{x \in S^{n-1} \mid \|x\|_T > 1/t\} \geq 1 - \left(\frac{C}{t}\right)^n .$$

Choosing t equal to, say, $10C$, we have a set $\mathfrak{A} \subset S^{n-1}$ of a very large measure (very high probability) where $\|x\|_T > 1/10C$. Then, using the standard argument of Asymptotic Theory, one finds some set of subspaces $A \subset G_{n,k}$, $k =]\frac{n}{2}[$, of, again, a large Haar probability measure (i.e. with a high probability) such that the unit euclidean sphere $S(E)$ of a subspace $E \in A$ is contained in some ε_0-neighborhood $\mathfrak{A}_{\varepsilon_0}$ of $\mathfrak{A}$ (we, of course, first fix $\varepsilon_0 > 0$ and then a number "10" above is, in fact, depends on this ε_0). This means precisely that for every $E \in A$, and some C_1 depending on C only

$$\frac{1}{C_1}|x| \leq \|x\|_T \qquad \text{for} \quad \forall x \in E$$

(also, for every $x \in \mathbb{R}^n$, $\|x\|_T \leq |x|$, because $D \subset T$). Therefore, $d(E \cap X_T, \ell_2^k) \leq C_1$.

(The reader who feels that he needs to fill in the details on the above outline proof, may consult the books [MSch]] or [P1].)

We continue now as in the proof of Theorem 2: there are two orthogonal subspaces E_1 and E_2, $E_1 \oplus E_2 = \mathbb{R}^n$ and $\dim E_1 =]\frac{n}{2}[$ (i.e. $\dim E_2 = [\frac{n}{2}]$) and such that both of them belong to the set A, i.e.

$$\frac{1}{C_1}|x| \leq \|x\| \qquad \text{for} \quad x \in E_1 \cup E_2 .$$

Starting from this inequality (which has exactly the same form as (2.4)) we construct, as in the proof of Theorem 2, an orthogonal operator $u_2 \in O_n$ such that for any $x \in \mathbb{R}^n$

$$(|x| \geq) \; \frac{\|x\|_T + \|u_2 x\|_T}{2} \geq \frac{1}{\sqrt{2}C_1}|x|$$

which proves the Theorem 1.B.

Remark. In the proof of both steps of Theorem 1.B, constructing an orthogonal operator u_1 in Theorem 4 (after choosing a position $\widehat{K} = vK$) and in the above construction of u_2, we did not deal with the Haar measure on O_n. We worked with special orthogonal operators (reflections) of the form $P_1 - P_2$ where P_i's are orthogonal projections onto pairwise orthogonal complemented subspaces. The probability measure which we induced on such operators came from the Haar probability measure on a suitable Grassmann manifold $G_{n,k}$. In this respect, we did not fulfill our promises in the remarks following Theorem 1. However, there is another approach which achieves it.

Let us quickly outline it. I intend this piece (up to the end of this section) for the experts.

Let $K \subset \mathbb{R}^n$ be in the main position (i.e. the euclidean ball $D \subset \mathbb{R}^n$ is an $\mathcal{M}_K$-ellipsoid). Then $\operatorname{Vol} D = \operatorname{Vol} K$ and for a universal constant C and any $T \subset \mathbb{R}^n$

$$\operatorname{Vol}(K + T) \le C^n \operatorname{Vol}(D + T)$$

(see [M2] or [P1] for all the facts used). Also D is an $\mathcal{M}$-ellipsoid of AK for any orthogonal rotation $A \in O_n$ which implies

$$\operatorname{Vol}(K + AK) \le (2C)^n \operatorname{Vol} D .$$

Define $K + \varepsilon D = K_\varepsilon$ (we fix some positive $0 < \varepsilon \le 1$). Then, again,

$$\operatorname{Vol} K_\varepsilon = C_0^n \operatorname{Vol} D$$

for $C_0 \le 2C$. Also $C_0 \cdot D$ is an $\mathcal{M}$-ellipsoid of K_ε and therefore there is a covering $\overline{\mathfrak{N}}$ of K_ε by D with the covering number $\#\overline{\mathfrak{N}} = N(K_\varepsilon, D) \le e^{an}$ for some universal, independent of n or K) a. The line of computation (2.10) shows that

$$\mu\{x \in S^{n-1} \mid \|x\|_{K_\varepsilon} \le 1/t\} \le \left(\frac{C_0}{t}\right)^n = e^{-n \log t/C_0} .$$

Take t such that $a < \log t/C_0$. Then any set $\mathfrak{N}$ of at most e^{an} points on S^{n-1} can be shifted (by one rotation $u \in O_n$) out of this small set, i.e. for every $x \in \mathfrak{N}$, $\|ux\|_{K_\varepsilon} > 1/t$. Take $\mathfrak{N} = \{\frac{x}{|x|} \mid x \in \overline{\mathfrak{N}}\}$. Note that there is a set $\mathfrak{A} \subset O_n$ of large (Haar probability) measure on O_n (exponentially close to 1) such that *any* $u \in \mathfrak{A}$ could do the job (I mean that $\|ux\|_{K_\varepsilon} > 1/t$ for any $x \in \mathfrak{N}$ and any $u \in \mathfrak{A}$). Consider the part $\overline{\mathfrak{N}}_t$ of the covering $\overline{\mathfrak{N}}$ which consists of all $x \in \overline{\mathfrak{N}}$, $|x|_{K_\varepsilon} > t$, i.e. the part which belongs to $K_\varepsilon \backslash tD$ and covers $K_\varepsilon \backslash (t+1)D$. Then, for every such $x \in \overline{\mathfrak{N}}_t$, $\frac{ux}{|ux|} \notin K_\varepsilon$ which means that euclidean ε-neighborhoods on the sphere

S^{n-1} of these points are *out of* K. It follows that D-neighbourhoods of points $x \in \overline{\mathfrak{N}}_t$ don't intersect K if $t > 1/\varepsilon$ (this is the balance to choose $\varepsilon > 0$ and t). This means that every point x of K, which has $|x| = t + 1$, will be out of K after rotation: $\|ux\|_K > 1$. So

$$\|x\|_K + \|ux\|_K > \frac{1}{t+1}|x| . \tag{2.11}$$

Therefore, by duality, for $u_1 = u^*$

$$T = K^\circ + u_1 K^\circ \supset \frac{1}{t+1} D \tag{2.12}$$

(for $u_1 \in O_n$ of very high probability). This is the first step of the proof. Note that if D is an $\mathcal{M}$-ellipsoid for K then it is an $\mathcal{M}$-ellipsoid for K° and (by Santaló's inequality) $\operatorname{Vol} K^\circ \le \operatorname{Vol} D$,

$$\operatorname{Vol}(K^\circ + u_1 K^\circ) \le C_1^n \operatorname{Vol} D$$

for a universal constant C_1. We repeat exactly the above reasoning starting with T and will come to the inequality (2.11): $\||x\|| = \|x\|_T + \|u_2 x\|_T > c|x|$ for some absolute constant $c > 0$. Also, by (2.12), the new norm $\|| \cdot \||$ is majorized from above by $(t+1)|x|$ and, this means, it is uniformly equivalent to the euclidean norm.

3. Applications

3a. Indicatrix of Growth of a Polynomial of Many Variables.

Consider a Loranian polynomial $P(z_1, \ldots, z_n) = \sum c_{k_1, \ldots, k_n} z_1^{k_1} \cdot \ldots \cdot z_n^{k_n}$ where $z_i \in \mathbb{R}$ and $k_i \in \mathbb{Z}$ (i.e. we also allow negative integer powers). We write, in brief, $P(\bar{z}) = \sum_{\bar{k}} c_{\bar{k}} \bar{z}^{\bar{k}}$, where, clearly $\bar{z}^{\bar{k}}$ means $z_1^{k_1} \cdot \ldots \cdot z_n^{k_n}$. Let $A(P)$ be the support of $P(\bar{z})$, i.e.

$$A(P) = \{\bar{k} \in \mathbb{Z}^n \mid c_{\bar{k}} \ne 0\} \subset \mathbb{Z}^n .$$

We will study a growth of such polynomial in different "exponential" directions $\bar{\lambda} \in \mathbb{R}^n$. This means that we consider a one-parametric family $\bar{z}_{\bar{\lambda}}(t) = t^{\bar{\lambda}} \equiv (t^{\lambda_1}, \ldots, t^{\lambda_n})$. Then we obtain $P_{\bar{\lambda}}(t) = P(t^{\bar{\lambda}}) = \sum_{\bar{k}} c_{\bar{k}} t^{(\bar{\lambda}, \bar{k})}$, where $(\bar{\lambda}, \bar{k}) = \sum_{i=1}^n \lambda_i k_i$ is the standard inner product in $\mathbb{R}^n$. We do not restrict ourselves to integer $\bar{\lambda}$'s and, therefore, we can obtain not only integer powers. We continue to call such "generalized" polynomials simply polynomials. Clearly, the growth of $P_{\bar{\lambda}}(t)$ is defined by the maximal term $P_{\bar{\lambda}}(t) = a t^{h(\bar{\lambda})} + \ldots$ where $a \ne 0$ and

$$h_P(\bar{\lambda}) = \max \left\{ \sum_1^n \lambda_i k_i \mid \bar{k} \in A(P) \right\} .$$

Therefore, it depends, in fact, on the set of all integer points of the set

$$N(P) = \text{Conv}\, A(P)$$

called the Newton polyhedron of P. We assume below that $0 \in N(P)$, and, moreover, 0 is an inner point of $N(P)$ (so, we assume that $N(P)$ is a body). The function $h_P(\overline{\lambda})$ is the support functional of $N(P)$ and defines the polar body of $N(P)$:

$$G(P) = [N(P)]^\circ = \{\overline{\lambda} \in \mathbf{R}^n \mid h_P(\overline{\lambda}) \le 1\}\,.$$

This body is called the "indicatrix of growth" of polynomial P. A straightforward interpretation of the duality relation (2.1) gives an interesting relation between a distribution of non-zero terms of polynomial P, i.e. $N(P)$, and the growth of P in different directions, i.e. $G(P)$. Note, that we apply (2.1) to, generally, non-centrally symmetric sets $N(P)$ and $G(P)$. In [M4], we specially emphasized (see Final Remark 3) that inequality (2.1) is also satisfied in the non-centrally symmetric case.

Corollary 3.1. *Fix* κ, $0 < \kappa < 1$ *and an integer* k, $1 \le k < (1-\kappa)n - C$ *where* C *is a universal constant from (2.1). Let, for some* $R > 0$, *any* k-*dimensional subspace* $E \in G_{n,k}$ *contains a point* $x \in E \cap N(P)$ *with the euclidean norm* $|x| > R$. *Then, there is an* $(n-k-\kappa n - C)$-*dimensional subspace* $F \subset \mathbf{R}^n$ *such that for any (direction)* $\overline{\lambda} \in F$ *the growth of* $P_{\overline{\lambda}}(t)$ *is at least* $t^{\kappa R \cdot |\overline{\lambda}|}$ *where* $|\overline{\lambda}| = \left(\sum \lambda_i^2\right)^{1/2}$ *is the euclidean norm of* $\overline{\lambda}$. *This means that* $\lim\limits_{t \to \infty} |P_{\overline{\lambda}}(t)|/t^{\kappa R \cdot |\overline{\lambda}|} > 0$.

The Low M^*-estimate (2.3) also has a similar interpretation. For any convex body $K \subset \mathbf{R}^n$ such that 0 belongs to the inner part $\overset{\circ}{K}$ of K consider the Minkowski functional $\varphi_K(x)$ of K. This is a positively homogeneous function: $\varphi_K(\lambda x) = \lambda \varphi_K(x)$ for $\lambda > 0$, such that $K = \{x \in \mathbf{R}^n \mid \varphi_K(x) \le 1\}$. Define, as in section 2,

$$M_K = \int\limits_{x \in S^{n-1}} \varphi_K(x) d\mu(x)\,,$$

where μ is the rotation invariant probability measure on S^{n-1}. Similarly, for any $x \in \overset{\circ}{K}$, we consider a set $K - x$ and define M_{K-x}. We call $x_0 \in K$ the M-center of K if

$$\min_x M_{K-x} = M_{K-x_0}\,.$$

(If K is centrally symmetric then 0 is the M-center of K.) We apply these notions to the set $N(P)$. Clearly, a shift $N(P) + x_0$ is the Newton polyhedron of a polynomial $P_1 = \overline{z}^{x_0} P(\overline{z})$

where, as before, $\bar{z}^{x_0} = z_1^{x_0^{(1)}} \cdot z_2^{x_0^{(2)}} \cdot \ldots \cdot z_n^{x_0^{(n)}}$ and $x_0^{(i)}$ is the i-th coordinate of $x_0 \in \mathbb{R}^n$. Of course, only if $x_0 \in \mathbb{Z}^n$ we obtain a usual polynomial P_1, but, for any $x_0 \in \mathbb{R}^n$, we can consider, in the same way, a general polynomial and its Newton polyhedron. The next statement is an immediate consequence of (2.3) and a measure estimate (see [M5]), which is a preliminary step in the proof of (2.3).

Define for a polynomial $P(\bar{z})$

$$M(P) = M_{N(P)} .$$

Note that a larger set $N(P)$ produces a smaller number $M(P)$.

Corollary 3.2.

a. Let, for $1 > \theta > 0$, $B_\theta = \{\bar{\lambda} \in S^{n-1} , \text{ i.e. } |\bar{\lambda}| = (\sum_1^n \lambda_i^2)^{1/2} = 1 \mid \frac{\theta}{M(P)} \leq h_P(\bar{\lambda})\}$. Then

$$\mu(B_\theta \subset S^{n-1}) \geq 1 - c\sqrt{n}(\sin\theta)^{n-2} .$$

b. For any θ, $0 < \theta < 1$, there is a $[\theta n - C]$-dimensional subspace $F \hookrightarrow \mathbb{R}^n$ (where C is a universal constant) such that for any direction $\bar{\lambda} \in F$, $|\bar{\lambda}| = 1$, the growth of $P_{\bar{\lambda}}(t)$ is at least

$$t^{\sqrt{1-\theta}/M(P)} . \tag{3.1}$$

Remarks. 1. We can first perform a shift of the set $N(P)$ and to use $M(P - x)$ for some $x \in N(P)$. Then, returning to describe the indicatrix of growth of the original polynomial P, instead of a low estimate (3.1),we obtain:

for any $x \in N(P)$ and for any θ, $0 < \theta < 1$, there is a $[\theta n - C]$-dimensional subspace $F \hookrightarrow \mathbb{R}^n$ such that for any $\bar{\lambda} \in F$, $|\bar{\lambda}| = 1$,

$$\lim_{t \to \infty} P_{\bar{\lambda}}(t) / t^{(x,\bar{\lambda}) + \sqrt{1-\theta}/M(P-x)} > 0 . \tag{3.2}$$

Note also, that to minimize $M(P - x)$, we can use x being the M-center of $N(P)$.

2. We may apply this result in the form of (3.2) also in the case when $0 \notin N(P)$ as was assumed before.

3.b Estimates on a number of integer points in a convex body K and some of its transforms.

The problems which we discuss in this section are related to classical facts on estimates on a number of integer points in a fixed convex centrally symmetric body K. The best known result in this direction — Minkowski's theorem — states that $\operatorname{Vol} K > 2^n$ implies K contains a non-zero integer point.

We change this question in a few directions.

Problem A. Consider a family of positions $\operatorname{SL}(K) = \{vK \mid v \in \operatorname{SL}_n\}$ of a given body K. Find a position $\check{K} = vK$ such that it, and all or most of its, orthogonal rotations $\{u(vK) \mid u \in O_n\}$ contain many integer points; estimate this number from below.

Problem B. Fix $0 < \lambda < 1$. Find a projection of K onto some coordinate subspace $E \hookrightarrow \mathbb{R}^n$, of dimension at least λn which contains as many as possible integer points. This will mean, naturally, that K contains many points with $[\lambda n]$ coordinates which are distinct ($[\lambda n]$-dimensional) integers.

Problem C. How many integer points of $\mathbb{R}^n$ can be represented for some fixed orthogonal transformation u, as $x + uy$ for $x \in K$, $y \in K$?

We will use results of the previous sections to give some answers to the above problems. In fact, we mix directions as outlined in Problems A,B and C.

Let D_n be the n-dimensional euclidean ball of radius 1. Define $d(t,n) = \#\{tD_n \cap \mathbb{Z}^n\}$, i.e $d(t,n)$ is the number of integer points in the euclidean ball of radius t. For a coordinate subspace E of dimension k, we denote $\mathbb{Z}^k(E) = E \cap \mathbb{Z}^n$. The function $d(t,n)$ was, of course, intensively studied (see, i.e. [W]). However, the emphasis was usually put on an asymptotic of $d(t,n)$ when $t \to \infty$ and n is fixed. Our use of this function is the other way round. We keep t much smaller than n (and actually smaller than $\sqrt{n}$) and often even bounded. At the same time, we are interested in an asymptotic for $n \to \infty$. Observe that, obviously

$$d(t,n) \geq \sum_{k=1}^{[t^2]} \binom{n}{k} \cdot 2^k \geq \binom{n}{[t^2]} 2^{[t^2]} . \tag{3.3}$$

Proposition 3.3. *Let $K \subset \mathbb{R}^n$ be a convex centrally symmetric compact body and $\operatorname{Vol} K = \operatorname{Vol} D_n$. There is a position $K_m = vK$, $v \in \operatorname{SL}_n$, such that for any λ, $0 < \lambda < 1$, and some set $\mathfrak{A}_\lambda \subset O_n$ of orthogonal rotations of a probability "almost" one (i.e. $\mu(\mathfrak{A}_\lambda) \geq 1 - \alpha(\lambda)^n$ for some $\alpha(\lambda)$, $0 < \alpha(\lambda) < 1$, and μ being the Haar probability measure on O_n) for any $u \in \mathfrak{A}_\lambda$,* *there is the orthogonal (coordinate) projection $P_{\lambda n}$ onto some coordinate $k = [\lambda n]$-dimensional subspace E and for any $t > 0$*

$$\#\{P_{\lambda n}(u(tK_m)) \cap Z^k(E)\} \geq d(t \cdot c(\lambda); k)$$

and

$$\#\{P_{\lambda n}(u(tK_m^\circ)) \cap Z^k(E)\} \geq d(t \cdot c(\lambda); k),$$

where $c(\lambda) \geq c(1 - \lambda)^\beta$ for a universal $c > 0$ and some $\beta > 0$ and K_m° is the polar body of K_m.

Proof is a straightforward consequence of some properties of the main position K_m of K. We introduced this position in section 2 but the properties which were mentioned there are not enough for this result. We use the following fact (originated in [M6],[M2]; see Pisier's book [P1]):

Let $\mathrm{Vol}\, K_m = \mathrm{Vol}\, D$; then for every λ, $0 < \lambda < 1$, there is a set $\mathcal{B}_\lambda \subset G_{n,k}$ for $k = [\lambda n]$-dimensional subspaces of a probability $\nu(\mathcal{B}_\lambda) \geq 1 - \alpha(\lambda)^n$ for some $0 < \alpha(\lambda) < 1$ (and ν here is the Haar probability measure on Grassmann manifold $G_{n,k}$) such that for every subspace $E \in \mathcal{B}_\lambda$ and the orthogonal projection P_E onto this subspace

$$P_E K_m \supset c(\lambda)(D \cap E)$$

where $c(\lambda) \geq c(1 - \lambda)^\beta$ for a universal $c > 0$ and some $\beta > 0$ (in my original proof β was not the best; the best estimate on $\beta \sim \frac{1}{2}$ follows from Pisier's proof – see his book [P1]). This is, of course, immediately translated into the first estimate of the Proposition. To obtain simultaneously the same estimate for the polar body, note that $(K_m)^\circ$ is, automatically, in its main position and, by Bourgain and the author result [BM] on the reverse Blaschke-Santaló inequality, $\mathrm{Vol}\, K_m \geq c_1^n \mathrm{Vol}\, D$ for some universal constant $c_1 > 0$ which is elaborated into a constant c which estimates $c(\lambda)$. $\quad\square$

In the next applications we use inequality (2.1') and Theorem 2 which allow us to leave unchanged the original position of body K. We defined $\bar{t}(r)$ at the start of section 2 as follows (we translate the definition into a geometric language of convex body K instead of normed space X):

$\bar{t}(r) \equiv \bar{t}(K; r) = \sup\{k \in \mathbb{N} \mid \exists \mathcal{B} \subset G_{n,k}$ of a probability $\nu(\mathcal{B}) > 1 - \frac{1}{n}$ such that for every subspace $E \in \mathcal{B}$, $(D_n \cap E) \cdot r \supset (K \cap E)\}$. In the dual form, it means that

$$\frac{1}{r}(D_n \cap E) \subset P_E K^\circ$$

where P_E is the orthogonal projection onto subspace E, $\dim E = \bar{t}(r)$. Similarly, there is a subspace F, $\dim F = \bar{t}(K^\circ; \rho) \equiv \bar{t}^*(\rho)$, such that

$$\frac{1}{\rho}(D_n \cap F) \subset P_F K \ .$$

By inequality (2.1'), if $\rho = \frac{1}{\kappa r}$ then $\bar{t}^*(\rho) \geq (1 - \kappa)n - \bar{t}(r) - o(n)$ and a subspace F is a "random" one in $G_{n, t^*(\rho)}$.

Joining this information with (3.3), we obtain the following corollary.

Proposition 3.4. *For any $K \subset \mathbf{R}^n$ as above, any $r > 0$ and κ, $0 < \kappa < 1$, there are integers k and k^*, $k + k^* \geq (1 - \kappa)n - o(n)$, where $o(n)/n \to 0$ for $n \to \infty$ and the function $o(n)$ does not depend on the choice of body K, such that for a fixed pair of coordinate subspaces E and F, $\dim E = k$ and $\dim F = k^*$, there is a set $\mathfrak{A} \subset O_n$ of a large measure and for any "random" rotation $u \in \mathfrak{A}$*

$$\#\{P_E(r(uK)^\circ) \cap Z^k(F)\} + \#\left\{P_F\left(\frac{1}{\kappa r}uK\right) \cap Z^{k^*}(F)\right\} \geq 2(1 - \kappa)n - o(n) \ ,$$

where P_E and P_F are the orthogonal projections onto E and F respectively.

In the next and last application we use Theorem 2 which states that $c < r_2(K) \cdot r_3(K^\circ)$ and $c < r_2(K^\circ)r_3(K)$ for some universal constant $c > \frac{1}{21}\sqrt{2/3} \simeq 0.03888$.

We use this fact in the following form; either one of $r_2(K)$ or $r_2(K^\circ)$ is at least c or both $r_3(K) \geq 1$ and $r_3(K^\circ) \geq 1$.

Proposition 3.5. *Let $K \subset \mathbf{R}^n$ be, as above, any convex, centrally symmetric, compact body and let K° denote its dual body. Then either there is an orthogonal operator $u \in SO_n$ such that for any $t \geq 1$ and a universal constant $c > 0.03888$*

$$\text{either} \quad \#\left\{z \in Z^n \ \middle| \ \exists x \in tK \text{ and } y \in tK \text{ such that } z = \frac{x + uy}{2}\right\} \geq d(ct, n)$$

$$\text{or} \quad \#\left\{z \in Z^n \ \middle| \ \exists x \in tK^\circ \text{ and } y \in tK^\circ \text{ such that } z = \frac{x + uy}{2}\right\} \geq d(ct, n) \tag{3.4}$$

or there are two orthogonal operators $u_1 \in SO_n$, $u_2 \in SO_n$ such that for any $t \geq 1$

$$\#\left\{z \in Z^n \ \middle| \ \exists x_1, x_2, x_3 \in tK \text{ and } z = \frac{x_1 + u_1 x_2 + u_2 x_3}{3}\right\} \geq d(t, n)$$

and

$$\#\left\{z \in Z^n \ \middle| \ \exists x_1, x_2, x_3 \in tK^\circ \text{ and } z = \frac{x_1 + u_1 x_2 + u_2 x_3}{2}\right\} \geq d(t, n)$$

Note, that in (3.4) $d(ct, n) \geq 2n$ for $t \geq 1/c$ and, in particular, for $t \geq 21\sqrt{3/2}$.

Also $\operatorname{Vol} K$ and $\operatorname{Vol} K^\circ$ can be of the order $\left(\frac{1}{\sqrt{n}}\right)^n$ together with volumes of $(K + uK)$, $(K^\circ + uK^\circ)$ or $\operatorname{Vol} \dfrac{K + u_1 K + u_2 K}{3}$ and $\operatorname{Vol} \dfrac{K^\circ + u_1 K^\circ + u_2 K^\circ}{3}$ and therefore, in general, we would be not able to use Minkowski Theorem to prove existence of non-zero integer point in these bodies up to values of $t \gtrsim \sqrt{n}$. However, in fact, we know this already starting $t \geq 21\sqrt{3/2}$.

Acknowledgement

This paper was written when the author was visiting at the Institute for Mathematics, ETH-Zurich. I would like to thank the Institute and Professor H. Jarchow of Zurich University for their hospitality.

References

[BLM]   J. Bourgain, J. Lindenstrauss, V. Milman, Minkowski sums and symmetrizations, GAFA-Seminar Notes '86-'87, Springer Lecture Notes in Math. v. 1317 (1988),44-66.

[BM]   J. Bourgain, V.D. Milman, New volume ratio properties of convex xymmetric bodies in $\mathbf{R}^n$, Inventiones Math. 88 (1987), 319-340; see also, On Mahler's conjecture on the volume of the convex symmetric body and its polar, preprint I.H.E.S., March 1985, and Sections euclidiennes et volume des corps convexes symétriques. C.R. Acad. Sci. Paris, A300 (1985), 435-438.

[G]   Y. Gordon, On Milman's inequality and random subspaces which escape through a mesh in $\mathbf{R}^n$. GAFA, Israel Funct. Analysis Seminar (86/87), Springer Lecture Notes 1317 (1988), 84-106.

[K]   B. Kašin, Sections of some finite dimensional sets and classes of smooth functions, Izv. Acad. Nauk SSSR 41 (1977), 334-351 (in Russian).

[M1]   V.D. Milman, An inverse form of the Brunn-Minkowski inequality with applications to local theory of normed spaces, C.R. Acad. Sc. Paris 302, Ser. 1, No. 1 (1986), 25-28.

[M2]   V.D. Milman, Isomorphic symmetrization and geometric inequalities, GAFA-Seminar Notes '86-'87, Springer Lecture Notes in Math. 1317 (1988), 107-131.

[M3]   V.D. Milman, Entropy point of view on some geometric inequalities, C.R. Acad. Sc. Paris 306 (1988), 611-615.

[M4]   V. Milman, Spectrum of a position of a convex body and linear duality relations, Israel Math. Conf. Proceedings (IMCP) 3, Festschrift in Honor of Professor I. Piatetsk Shapiro, Weizmann Science Press of Israel 1990.

[M5]   V. Milman, Random subspaces of proportional dimension of finite dimensional normed spaces; approach through the isoperimetric inequality, Banach spaces, Proceedings, Missouri, 1984, Springer-Verlag, Lecture Notes in Mathematics 1166, 106-115 (1985).

[M6]   V.D. Milman, Geometrical inequalities and mixed volumes in Local Theory of Banach Spaces, Asterisque (special issue dedicated to Professor Laurent Schwartz) 131, 373-400.

[MSch]   V.D. Milman, G. Schechtman, Asymptotic theory of finite dimensional normed spaces, Springer-Verlag, Lecture Notes in Mathematics 1200, 156pp. (1986).

[P1]   G. Pisier, The Volume of Convex Bodies and Banach Space Geometry, Cambridge Tracts in Mathematics 94, Cambridge University Press, Cambridge (1989).

[P2]   G. Pisier, A new approach to several results of V. Milman, Journal für die Reine and Angew. Math. 393 (1989), 115-131.

[PT]   A. Pajor, N. Tomczak-Jaegermann, Subspaces of small codimension of finite dimensional Banach spaces, Proc. Amer. Math. Soc. 97 (1986), 637-642.

[ST]   S. Szarek, N. Tomczak-Jaegermann, On nearly Euclidean decompositions of some classes of Banach spaces, Compositio Math. 40 (1980), 367-385.

[W]   A. Walfisz, Gitterpunkte in mehrdimensionalen Kugeln, Warszawa, 1975.

[WSW]   W.D. Wallis, A.P. Street, J.S. Wallis, Combinatorics: Room Squares, Sum-free Sets, Hadamard Matrices, Lecture Notes in Mathematics 292, Springer-Verlag 1972.

MATHEMATICAL PROBLEMS IN THE THEORY OF QUANTUM CHAOS

Ya.G. Sinai

Landau Institute for Theoretical Physics
Academy of Sciences
Moscow, USSR

§1 Introduction

Theory of quantum chaos has both experimental and theoretical aspects. It contains in its title the word "chaos", not because of some erratic temporal behaviour of a quantum system, but because of its analysis of the role of classical chaos to properties of quantum systems. Theoretical works on quantum chaos are split into two parts. In one part people start with a classical chaotic dynamic system and try to quantize it. In contrast to this, in the other part everything starts with a quantum system and the problem is to try to find properties which are determined by the properties of the corresponding classical limit. It is natural that in the last case the methods are based upon the theory of quasi-classical approximation, and, in particular, on the precision of this approximation.

The main researchers in quantum chaos are M. Berry, I. Percival, M. Gutzwiller, J. Ford, A. Voros, G. Casati, B. Chirikov, G. Zaslavski, F. Izraelev, D. Shepeljanski, G. Berman. Since I am a beginner in this field I include only the names of those whose works have influenced my studies in this topic. One of the first works in quantum chaos was the paper by A. Einstein (1917). The main works in the theory of quasi-classical approximations include the well-known papers by J. Keller, R. Balyan and C. Bloch, V.P. Maslov. I was very glad to meet Professor U. Smilansky of the Weizmann Institute during my stay in Tel Aviv, and to have several very interesting discussions with him concerning the works done by him and his colleagues on quantum chaos.

Mostly, we shall deal in these lectures with two-dimensional quantum systems. The corresponding models describe quantum particles on two-dimensional surfaces. Mathematically this means that we shall consider two-dimensional closed compact Riemannian surfaces Q equipped with some Riemannian metrics ds^2. The C^2-smoothness of the metrics is sufficient for our goals.

It has just been brought to our attention by the author that he recently submitted this paper also to the American Institute of Physics for publication and it has just appeared in "Chaos/XAOC" 1990, edited by D.K. Campbell. We thank the AIP for permission to go ahead with this publication.

The metrics ds^2 generates the Laplace operator Δ and the stationary states of the quantum particle are described by eigen-functions of Δ, i.e. by the solutions of the equation

$$-\Delta\psi_k = E_k\psi_k .$$

Here E_k are eigen-values of $-\Delta$ which are labelled in the increasing order:

$$0 < E_1 \leq E_2 \leq \ldots \leq E_k \leq \ldots .$$

Generically only strict inequalities hold.

We introduce the function $N(x)$ equal to the number of E_k which are less than x. It is the famous Weyl theorem which says that asymptotically

$$N(x) = \frac{\text{Area}(Q)}{4\pi}x + o(x) .$$

The detailed study of the asymptotic expansion of $o(x)$ is sometimes a very complex mathematical problem.

One of the main ideas of quantum chaos is that even for regular surfaces the function $N(x)$ behaves on small distances as a random function. There are at least two possible ways of describing this randomness.

1. Fix an interval (α,β), $0 < \alpha < \beta < \infty$, and introduce $\mu_x((a,b)) = \frac{1}{N(x)}\#\{E_k < x,\ E_k - E_{k-1} \in (\alpha,\beta)\}$. Then for every x we have a probability measure on R^1. The Weyl asymptotics easily implies that the family of measures μ_x is weakly compact. The weak limits of μ_x describe the limiting distributions of spacings between the nearest eigen-values.

2. Take $c > 0$ and for any integer $k \geq 0$ introduce the set $A_k(x) = \{E \leq x \mid (E, E+c)$ has k eigen-values$\}$. Then the probabilities $\pi_x(k) = \frac{\ell(A_k(x))}{x}$ where ℓ is the Lebesgue measure and their limiting points characterize the clustering properties of E_k.

Many qualitative arguments and numerical studies show that weak limits of μ_x, and $\pi_x = \{\pi_x(k)\}$ exist as $x \to \infty$. We consider these limiting distributions as describing statistical properties of $N(x)$ on scales of order of unity or the local randomness of $N(x)$. Mathematically it is a difficult problem to show the existence of weak limits because we have no convenient functional representations for the introduced probabilities. Moreover, for some special cases it might even happen that the limits really are not unique and there are several limit points.

The main statement in the theory of quantum chaos which was expressed explicitly by I. Percival, M. Berry, G. Zaslavsky, B. Chirikov and maybe by some other researchers is that the form and the properties of these limiting distributions are determined completely by the ergodic properties of the corresponding geodesic flow acting on the unit tangent bundle over Q generated by the metrics ds^2.

The paper by M. Berry and Tabor [1] contains very convincing arguments in favor of a very striking statement which says that if the geodesic flow is completely integrable than $w - \lim_{z \to \infty} \mu_z = \mu$ is an exponential distribution, i.e. $\mu(\alpha, \beta) = \int_\alpha^\beta \rho e^{-\rho u} du$ where ρ can be found from the Weyl asymptotics, and $w - \lim_{z \to \infty} \pi_z = \pi$ is Poisson, i.e. $\pi(\kappa) = e^{c\rho} \frac{(c\rho)^\kappa}{\kappa!}$. The density of the exponential distribution $\rho e^{-\mu u}$ remains positive as $u \to 0$. This property is interpreted in the physical literature as the absence of repulsion of levels because the probability of having two close eigen-values is approximately the same as the probability of having two neighboring eigen-values at any distance of order of unity.

It is widely believed that if the geodesic flow is ergodic, mixing, K-flow, etc., then the density of the limiting distribution μ behaves as $\operatorname{Const} u^\gamma$ as $u \to 0$ which is interpreted as a repulsion of levels. This type of behavior is known for the distribution of spacing between levels in various ensembles of matrices of large dimension which were studied by Wigner, Dyson, Mehta and others. Apparently it is the most interesting part of the theory of quantum chaos but unfortunately at the moment I have no definite results here and therefore shall not discuss it any more.

§2 Surfaces of Revolution, Their Laplace Operators and Quasi-classical Expressions of Eigen-values

In this and the next sections we shall discuss the above-formulated statement by Berry and Tabor. The simplest example of an integrable Hamiltonian system with two-degrees of freedom is the geodesic flow on a two-dimensional surface of revolution. Its integrability follows from the extra symmetry of the surface and the existence of the corresponding Clairot integral. We shall deal with the surfaces having the topology of the torus. Such surfaces are naturally described by continuous positive periodic functions f of some period h, $f(0) = f(h)$. The properties of smoothness of f will be clear from the context. The whole surface Q is the result of the rotation of the graph of f along the r-axis and the subsequent gluing of boundary circles (see Fig. 1).

For some technical reasons which we will discuss later we shall assume that f has only two non-degenerate critical points, being strictly monotone outside them. The natural coordinates on Q are r, φ, where φ is the angle counted from some fixed direction. The Riemannian metrics

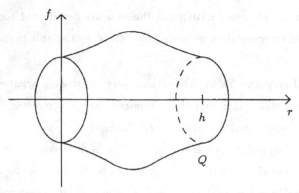

Fig. 1

in these coordinates has the form

$$ds^2 = f^2(r)d\varphi + \left(1 + \left(f'(r)\right)^2\right)dr^2 \ .$$

Now we can write down the explicit expression for the Laplace operator

$$\Delta\psi = \frac{1}{f\sqrt{1+(f')^2}} \frac{\partial}{\partial r}\left(\frac{f}{\sqrt{1+(f')^2}}\frac{\partial}{\partial r}\psi\right) + \frac{1}{f^2}\frac{\partial^2\psi}{\partial\varphi^2} \ .$$

Look for the eigen-functions in the form $\psi_{n,m} = e^{2\pi in\varphi}\cdot v_{n,m}(r)$. This representations follows from the symmetry of the metrics which is preserved on the quantum level as some symmetry of the Laplacian. For $v_{n,m}$ we get the equation

$$\frac{1}{f\sqrt{1+(f')^2}}\frac{d}{dr}\frac{f}{\sqrt{1+(f')^2}}\frac{d}{dr}v_{n,m} - \frac{4\pi^2 n^2}{f^2}v_{n,m} = -E_{n,m}v_{n,m}$$

where $E_{n,m}$ is the corresponding eigen-value. Introduce the new variable s through the formula

$$\frac{dr}{ds} = \frac{f}{\sqrt{1+(f')^2}} \ ,$$

or $s = \int_0^r \frac{1}{f(t)}\sqrt{1+(f't)^2}\,dt$. Then f becomes the periodic function of period $h_1 = \int_0^h \frac{1}{f(t)}\sqrt{1+\left(f'(t)\right)^2}\,dt$ and the equation for $v_{n,m}$ takes its final form

$$\frac{d^2}{ds^2}v_{n,m} + \left(E_{n,m}f^2 - 4\pi^2 n^2\right)v_{n,m} = 0 \ .$$

We shall study large eigen-values $E_{n,m}$. Take a parameter L which will later tend to infinity. Assume that n is of order L. To be more precise, fix two positive numbers a_1 and a_2 and consider $a_1 L \leq n \leq a_2 L$. The eigen-value $E_{n,m}$ will be of order L^2. In these circumstances we can use the quasi-classical approximation for the eigen-functions $v_{n,m}$ and write them in the form $v_{n,m} = \exp\left\{iL\left(\sigma_{n,m}^{(0)} + \frac{1}{iL}\sigma_{n,m}^{(1)} + \ldots\right)\right\}$ where the dots mean terms of a smaller order. For $\sigma_{n,m}^{(0)}$ we have the equation

$$\left((\sigma_{n,m}^{(0)})'\right)^2 = (\varepsilon_{n,m}f^2 - 4\pi^2\nu_n^2)$$

where we put $\varepsilon_{n,m} = E_{n,m}L^{-2}$, $\nu_n = nL^{-1}$, or

$$\sigma_{n,m}^{(0)} = \pm \int \sqrt{\varepsilon_{n,m}f^2 - 4\pi^2\nu_n^2}\, dt\ . \tag{1}$$

The interpretation of the square root in (1) is well-known (see, e.g. [2]). Fix ν_n and consider for any ε the domain $U(\varepsilon, \nu_n)$ where $U(\varepsilon, \nu) = \{x \mid \varepsilon f^2 - 4\pi^2\nu_n^2 > 0\}$. Due to our assumptions about f the domain $U(\varepsilon, \nu_n)$ consists of a single interval (see Fig. 2). It is the domain of a possible motion of the corresponding classical particle. The quasi-classical expressions give strongly oscillating

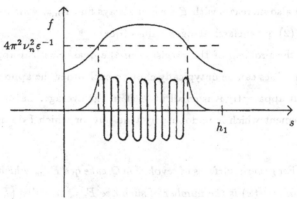

Fig. 2

expressions on $U(\varepsilon, \nu_n)$ which decay exponentially outside $U(\varepsilon, \nu_n)$. If f could have several critical points then $U(\varepsilon, \nu_n)$ might be more complicated. For some values of ε where the number of components changes, the structure of $U(\varepsilon, \nu_n)$ has singularities. In the vicinity of these values, quantum resonances take place and the formulas of the quasi-classical approximation become more complicated. We shall not discuss this here.

For the next term $\sigma_{n,m}^{(1)}$ we have the equation

$$(\sigma_{n,m}^{(1)})' = -\frac{1}{2}\left(\frac{(\sigma^{(0)})''}{(\sigma^{(0)})'}\right)$$

which gives $\sigma_{n,m}^{(1)} = -\frac{1}{2}\ln(\sigma_{n,m}^{(0)})' = -\frac{1}{4}\ln(\varepsilon^2 f^2 - 4\pi^2\nu^2)$. Now we can write down the usual quasi-classical expression for the eigen-functions

$$v_{n,m} = \frac{C_1}{\sqrt{\varepsilon f^2 - 4\pi^2\nu^2}} e^{iL\int\sqrt{\varepsilon f^2 - 4\pi^2\nu^2}\,dt} + \frac{C_2}{\sqrt{\varepsilon f^2 - 4\pi^2\nu^2}} e^{-iL\int\sqrt{\varepsilon^2 f^2 - 4\pi^2\nu^2}\,dt} .$$

The integration starts at the left end-point of $U(\varepsilon, \nu_n)$ or at the left solution of the equation $\varepsilon f^2 = 4\pi^2\nu^2$. It is worth mentioning that these formulas work successfully far from the boundary of $U(\varepsilon, \nu_n)$. Now we write down the famous quantization Bohr-Sommerfeld rules which give quasi-classical expressions for the eigen-values

$$\int_{s_1}^{s_2} \sqrt{(E_{n,m}f^2(s) - 4\pi^2 n^2)}\,ds = \pi\left(m + \tfrac{1}{2}\right) . \tag{2}$$

Here $s_1 = s_1(E_{m,n})$, $s_2 = s_2(E_{m,n})$ are the end-points of $U(\varepsilon_{m,n}, \nu_n)$. Equation (2) should be understood in the following way. Fix n, m of order L and start to increase E. Since the left-hand part of (2) also increases with E we can always find $E_{n,m}$ which solves (2). We shall call the solutions of (2) quasi-classical eigen-values (qce).

Discuss briefly the problem of the precision of the quasi-classical approximation. Since sometimes the eigen-values can be untypically close to each other the approximation cannot be uniformly small. But apparently it is small at least in the average. Let us formulate a quasi-theorem, i.e. a statement which is undoubtedly true but for which I do not have a complete proof at the moment.

Quasi-Theorem. *For generic surfaces of revolution Q take qce $E_{n,m}$ which are less than x and fix any $\varepsilon > 0$. Then if $N^{(1)}(x)$ is the number of such qce $E_{n,m}$ that dist $\left(E_{n,m}, \mathrm{spec}(-\Delta)\right) \leq \varepsilon$ then $\frac{N^{(1)}(x)}{N(x)} \to 1$ as $x \to \infty$.*

Apparently a stronger statement is valid which describes in a more precise way dist $\left(E_{n,m}, \mathrm{spec}(-\Delta)\right)$. Anyway further we shall deal only with qce. Also we neglect the term $\frac{1}{2}\pi$ in the right-hand part of (2). Later this error can be easily taken into account. Consider the functional equation

$$\int_{s_1(r)}^{s_2(r)} \sqrt{(r^2 f^2(s) - 4\pi^2 \sin^2\alpha)}\,ds = \pi\cos\alpha , \tag{3}$$

which determines r as an implicit function of α in the domain $\cos\alpha > 0$. Here $s_1(r), s_2(r)$ are the roots of the equation $r^2 f^2(s) = 4\pi^2 \sin^2\alpha$. If $r^2 f^2(s) \geq 4\pi^2 \sin^2\alpha$ everywhere then

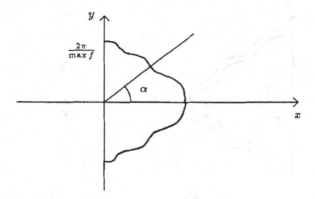

Fig. 3

$s_1(r) = 0$, $s_2(r) = h_1$. A typical form of the function $r = F(\alpha)$, is presented in Fig. 3. It is important also that it is symmetric with respect to the x-axis.

For any $(m, n) \in \mathbb{Z}^2$ put $\sin \alpha = \frac{n}{\sqrt{m^2+n^2}}$, $\cos \alpha = \frac{m}{\sqrt{m^2+n^2}}$, $\alpha = \alpha(m, n)$. Take two positive numbers E, c, and consider the inequality

$$E \leq E_{n,m} \leq E + c .$$

Further, $E_{n,m}$ will be of order L^2. Since we want to consider terms of order L it is more convenient to rewrite these inequalities in another form

$$\sqrt{E} \leq \sqrt{E_{n,m}} \leq \sqrt{E+c} = \sqrt{E} + \frac{c}{2\sqrt{E}} + \cdots \tag{4}$$

The dots mean as before terms of smaller order. We shall neglect them as well and as before the error can easily be taken into account. Since $E_{m,n}$ are the solutions of

$$\int_{s_1}^{s_2} \sqrt{E_{m,n} f^2(s) - 4\pi^2 n^2} \, ds = \pi m , \tag{2'}$$

i.e. $E_{m,n} = F(\alpha(m, n))\sqrt{m^2 + n^2}$, we have

$$\sqrt{E} \leq \sqrt{m^2 + n^2} F(\alpha(m, n)) \leq \sqrt{E} + \frac{c}{2\sqrt{E}}$$

or

$$\frac{\sqrt{E}}{F(\alpha(m, n))} \leq \sqrt{m^2 + n^2} \leq \left(\sqrt{E} + \frac{c}{2\sqrt{E}}\right) \cdot \frac{1}{F(\alpha(m, n))} . \tag{5}$$

We shall see that (5) have a beautiful geometric interpretation. Put $G_1(\alpha) = 1/F(\alpha)$ and consider for any $R \geq 1$ the curve γ_R whose equation in the polar coordinates takes the form

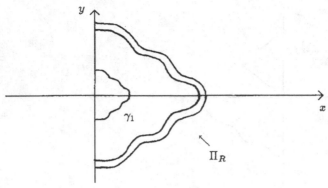

Fig. 4

$r = RG_1(\alpha)$. Denote by Π_R the closed curvilinear strip bounded by the curves $\gamma_{R+\frac{c}{2R}}$ and γ_R (see Fig. 4). Then the number of solutions of (5) is equal exactly to the number of points of the lattice Z^2 belonging to the strip $\Pi_{\sqrt{E}}$. It follows from the symmetry of γ_1 that this number is always even.

Now we can formulate differently the problem of distribution of spacings between qce. Take $L \to \infty$ and fix two positive numbers $a_1, a_2, a_1 < a_2$. Consider the solutions of (2′) such that $a_1^2 L^2 \le E_{n,m} \le a_2^2 L^2$. Fix also a number $c > 0$ and put $\xi(E, G_1)$ equal to the number of $E_{n,m} \in [E, E + c]$. The previous discussion shows it is equal to the number $\eta(\sqrt{E}, G_1)$ of points of the lattice Z^2 belonging to $\Pi_{\sqrt{E}}$. Denote $\mathcal{P}_\kappa(L; G_1)$ the Lebesgue measure of such $R = \sqrt{E}$, $a_1 L \le R \le a_2 L$, for which $\eta(R, G_1) = 2k$ and $p_\kappa(L, G_1) = \frac{1}{L(a_2 - a_1)} \mathcal{P}_\kappa(L; G_1)$.

Individual Poisson distribution. $\lim\limits_{L \to \infty} p_\kappa(L, G_1) = e^{-\lambda} \frac{\lambda^\kappa}{\kappa!}$ for some parameter $\lambda = \lambda(G_1)$. It is easy to see that in our notations $\lambda(G_1) = \frac{c}{2} \int G_1^2(\alpha) d\alpha$.

Averaged Poisson distribution. Assume that a probability distribution Prob on the Borel σ-algebra of the space of smooth curves γ_1 is given. Then

$$\lim_{L \to \infty} \int p_\kappa(L, G_1) d\operatorname{Prob}(G_1) = \int e^{-\lambda(G_1)} \frac{(\lambda(G_1))^\kappa}{\kappa!} d\operatorname{Prob}(G_1) \ .$$

In the next section we discuss the conditions under which we can prove the Averaged Poisson distribution and some version of the individual Poisson distribution.

Problems.

1. To prove the quasi-theorem.

2. To study the case of functions f having several critical points.

3. To generalize the described construction to the case of Liouville integrable metrics.

4. To study the quantum analogy of the KAM-theorem in the following form: for small generic perturbations of metrics of revolution a big fraction of eigen-values can be obtained with the help of the usual Bohr-Sommerfeld quantization rules.

§3 Poisson Distribution

Let us recall how the Poisson distribution appears in the traditional probability theory. One considers N independent random variables $\xi_1, \ldots, \xi_N$ taking the values 0 and 1 and $p_N = \text{Prob}\{\xi_\kappa = 1\} \sim \frac{\rho}{N}$ as $N \to \infty$. For the characteristic function $u_N(\lambda) = E e^{i\lambda\zeta_N}$, $\zeta_N = \xi_1 + \ldots + \xi_N$ we have the expression $u_N(\lambda) = \left(1 + \frac{\rho}{N}(e^{i\lambda} - 1) + o(N)\right)^N$ which tends to $\exp\{\rho(e^{i\lambda} - 1)\}$ which is the characteristic function of the Poisson distribution. A direct way to extend this derivation to dependent random variables is to write $u_N(\lambda)$ in the form

$$u_N(\lambda) = E e^{i\lambda\zeta_N} = E \prod_{\kappa=1}^{N} \left(1 + (e^{i\lambda\xi_\kappa} - 1)\right) =$$

$$= \sum_{r=0}^{N} (e^{i\lambda} - 1)^r \sum_{\{k_1, \ldots, k_r\}} \text{Prob}\{\xi_{k_1} = \ldots = \xi_{k_r} = 1\} \,.$$

and impose some conditions which permit replacing $\sum_{\{k_1, \ldots, k_r\}} \text{Prob}\{\xi_1 = \ldots = \xi_{k_r} = 1\}$ by $C_N^r \left(\frac{\rho}{N}\right)^r +$ some corrections. Following this direction we must assume that $\text{Prob}\{\xi_{k_1} = \ldots = \xi_{k_r} = 1\}$ behave as $\left(\frac{\rho}{N}\right)^r$. We shall see that in our problems it is not the case and the reasons for the Poisson distribution are in a sense quite different.

Since $G_1(\alpha)$ are even functions of α we shall consider only the angles $0 \le \alpha \le \frac{\pi}{2}$. Starting with this moment we consider a probability distribution P on the space $\mathcal{G}$ of positive functions $G_1(\alpha)$, $o \le \alpha \le \frac{\pi}{2}$. We list below the conditions which we have to assume about P.

1°. There exist positive constants C_1, C_2, C_3 such that with P – probability 1

$$C_1 \le G_1(\alpha) \le C_2 \,, \qquad \left|G_1'(\alpha)\right| \le C_3 \,.$$

2°. For any r and for any $0 < \alpha_1 < \alpha_2 < \ldots < \alpha_r < \frac{\pi}{2}$ there exists a density $\pi_r(y_r, \ldots, y_1 \mid \alpha_r, \alpha_{r-1}, \ldots, \alpha_1)$ of the joint probability distribution of the random variables $G_1(\alpha_1), G_1(\alpha_2), \ldots, G_1(\alpha_r)$ which is the C^1-function of its variables.

3°. The conditional density $\pi_{r+1}(y_{r+1} \mid G_1(\alpha_r) = y_r, \ldots, G_1(\alpha_1) = y_1; \alpha_r, \ldots, \alpha_1)$ satisfies the inequality

$$\pi_{r+1}\left(y_{r+1} \mid G_1(\alpha_r) = y_r, \ldots, G_1(\alpha_1) = y; \alpha_r, \ldots, \alpha_1\right) \le \frac{C_4}{\ell}$$

for a constant C_4, where ℓ is the interval of possible values of y_{r+1}.

Condition 1° can apparently be weakened. Condition 2° is essential. It shows that G_1 depends on infinitely many random parameters. It also means that we cannot consider the case of G_1 being polynomials of a bounded degree. Condition 3° means that G_1' is more or less uniformly distributed and G_1 has now preferable directions. It would be logically more consistent to consider a probability distribution on the space of functions f and to derive the needed properties of the distribution of G_1 but in my opinion the transition from the probability distribution on f to the probability distribution on G_1 is a technical problem.

Return now to our situation. For any positive function G_1 and any R we consider the strip Π_R bounded by the curves γ_R and $\gamma_{R+\frac{c}{2R}}$ where γ_R is described by the equation $r = RG_1(\alpha)$ in the polar coordinates. The value of the parameter R belongs to the interval $(a_1 L, a_2 L)$ where a_1, a_2 are two fixed numbers and $L \to \infty$. We consider the random variable $\eta(R, G_1) = \#(\Pi_R \cap \mathbb{Z}^2)$ which is a random variable defined on the direct product $(a_1 L, a_2 L) \times \mathcal{G}$. The measure on $[a_1 L, a_2 L]$ is the normalized Lebesgue measure. Denote $p_L(n) = \text{Prob}\{\eta(R, G_1) = n\}$.

Take $(m, n) \in \mathbb{Z}^2$ and introduce the interval $\mathcal{D}_{m,n}$ of such values of R for which $(m, n) \in \Pi_R$. It is easy to calculate that up to terms of a higher order of smallness the center of $\mathcal{D}_{m,n}$ is $d_{m,n} = \frac{\sqrt{m^2+n^2}}{G_1(\alpha(m,m))}$ and the length $\ell(\mathcal{D}_{m,n}) = \ell_{m,n} = \frac{c}{2\sqrt{m^2+n^2}}G(\alpha(m,n))$. Thus on the interval $(a_1 L, a_2 L)$ we have intervals $\mathcal{D}_{m,n}$ whose length is of order L^{-1} and the total number of such intervals is of order L^2. Thus a mean number of intervals covering a point is finite. It is easy to see that $\eta(R, G_1)$ equals exactly to the number of intervals $\mathcal{D}_{m,n}$ covering R. Introduce the random variable $\chi_{(m,n)}(R, G_1)$ which is equal to 1 if $R \in \mathcal{D}_{m,n}$ and 0 otherwise. Thus $\eta(R, G_1) = \sum_{(m,n)\in\mathbb{Z}^2} \chi_{(m,n)}(R, G_1)$ and we are in a situation typical for the Poisson limit theorem since $E\chi_{(m,n)}(R, G_1) = \mathcal{O}(L^{-2})$ and the number of summands in the last sum is $\mathcal{O}(L^2)$. However the product $\chi_{(m_1,n_1)}(R, G_1) \cdot \ldots \cdot \chi_{(m_r,n_r)}(R, G_1)$ takes the value 1 on the interval $\bigcap_{j=1}^{r} \mathcal{D}_{(m_j,n_j)}$ and is 0 otherwise. Typically the length $\ell(\bigcap_{j=1}^{r} \mathcal{D}_{(m_j,n_j)})$ is either 0 or of order L^{-1} and thus the direct generalization of the usual proof of the Poisson limit theorem does not work.

We are now going to discuss the proof of the averaged Poisson distribution. The precise formulation of the theorem is the following.

Theorem 1. *Under the conditions 1-3 there exists a weak limit of the distributions of the random variables $\eta(R, G_1)$ as $L \to \infty$ which is the mixture of the Poisson distributions with the parameter $\frac{c}{2} \int_0^{\pi/2} G_1^2(\alpha)d\alpha$.*

The proof is based upon the following simple lemma. Let ξ be a random variable taking non-negative integer values. For any $k \geq 0$ introduce the new random variable $\xi^{(k)}$ which is equal to zero if $\xi < k$ and $\xi^{(k)} = C_\xi^k$ otherwise.

Lemma 1. *Random variable ξ has the Poisson distribution with parameter ρ iff $E\xi^{(k)} = \frac{1}{k!}\rho^k$ for any k. Random variable ξ has the distribution which is a mixture of Poisson distributions iff $E\xi^{(k)} = \frac{1}{k!}\int \rho^k dS(\rho)$ for any k where S is the probability distribution for ρ.*

In view of the lemma, our goal is to show that for any k

$$\lim_{L \to \infty} \sum p_L(n) \cdot C_n^k = \frac{c^\kappa}{2^\kappa k!} E\left(\int G_1^2(\alpha) d\alpha \right)^k =$$

$$= \frac{c^k}{2^k k!} \int \cdots \int \pi_\kappa(y_1, \ldots, y_\kappa \mid \alpha_1, \ldots, \alpha_\kappa) y_1^2 \cdots y_\kappa^2 \prod_{j=1}^{k} dy_j \, d\alpha_j \, . \quad (6)$$

In order to prove (6) we shall study the statistics of mutual intersections of different intervals $\mathcal{D}_{m,n}$. Fix an interval B on the R-axis.

Definition 1. The intervals $\mathcal{D}_{m',n'}$ and $\mathcal{D}_{m'',n''}$ have B-intersection if $d_{m',n'} - d_{m'',n''} \in B$.

We shall consider intervals B of the form $B = B^{(s)} = [\frac{s}{L}\delta, \frac{s+1}{L}\delta]$ where δ is a small constant. Take now k intervals $\Delta^{(j_1)} < \Delta^{(j_2)} < \ldots < \Delta^{(j_\kappa)}$, $\Delta^{(j_t)} \subset [0, \frac{\pi}{2}]$, $1 \leq t \leq k$, and $(k-1)$ intervals $B_2, \ldots, B_k$, each $B_j = B^{(s_j)}$. Let $N_{(m_1,n_1)} = N_{(m_1,n_1)}(\Delta^{(j_1)}, \ldots, \Delta^{(j_k)}; B_2, \ldots, B_\kappa \mid G_1)$ be the number of k-tuples $(m_1, n_1), \ldots, (m_k, n_k)$ such that

1) $\alpha(m_t, n_t) \in \Delta^{j_t)}$, $t = 1, \ldots, k$;

2) $\mathcal{D}_{(m_t,n_t)}$ and $\mathcal{D}_{(m_1,n_1)}$ have B_t-intersection, $t = 2, \ldots, k$.

Lemma 2.

$$EN_{(m_1,n_1)} = \int \cdots \int dy_1 dz_2 \ldots dz_k \cdot z_2^2 \cdot \ldots \cdot z_\kappa^2 \left(\frac{\sqrt{m_1^2 + n_1^2}}{y_1 L} \right)^{k-1}$$

$$\cdot \pi_k\left(y_1, z_2, \ldots, z_k \mid \alpha(m_2, n_1), \overline{\alpha}^{(2)}, \ldots, \overline{\alpha}^{(k)}\right) \cdot \prod_{s=2}^{k} \ell(\Delta^{(j_s)}) \delta^{k-1}(1 + \gamma_1) \, .$$

Here γ_1 is a remainder term which tends to zero as both $\delta \to 0$ and $\ell(\Delta^{(j_s)}) \to 0$ provided that all distances $\text{dist}(\Delta^{(j_{s_1})}, \Delta^{(j_{s_2})}) \geq \beta$ where β is an arbitrary constant.

The proof of the lemma consists of direct calculations.

Definition 2. A set of $(k-1)$ intervals $\{B_2, \ldots, B_k\}$ is essential if for any possible positions of the intervals $\mathcal{D}_{(m_j, n_j)}$, $j = 2, \ldots, k$, having B_j-intersection with $\mathcal{D}_{(m_1, n_1)}$ the length $\ell\left(\bigcap_{j=2}^{k} \mathcal{D}_{(m_j, n_j)}\right) > 0$.

The variance of this length is not more than const $k\delta \cdot L^{-1}$ while $\ell(\mathcal{D}_{(m_j, n_j)}) = \mathcal{O}(L^{-1})$. Denote

$$\ell(B_2, \ldots, B_k) = \frac{\max \ell\left(\bigcap_{j=1}^{k} \mathcal{D}_{(m_j, n_j)}\right) + \min \ell\left(\bigcap_{j=1}^{k} \mathcal{D}_{(m_j, n_j)}\right)}{2}$$

where max and min are taken over all possible positions of $\mathcal{D}_{(m_j, n_j)}$ having B_j-intersection with $\mathcal{D}_{(m_1, n_1)}$, $j = 2, \ldots, k$. The notation $\sum^{(e)}$ means further the summation over all essential $(k-1)$-tuples $\{B_2, \ldots, B_k\}$.

Lemma 3. $\sum^{(e)} \ell(B_2, \ldots, B_k) \cdot \delta^{k-1} = L^{-1}\left(\left(\frac{cL}{2\sqrt{m_1^2 + n_1^2}} y_1\right)^k + \mathcal{O}(\delta)\right).$

Now we describe the last step in the proof of Theorem 1. Let $\chi_{(m_1, n_1), \ldots, (m_k, n_k)}(G_1; B_2, \ldots, B_k)$ be equal to 1 if $\mathcal{D}_{(m_j, n_j)}$ and $\mathcal{D}_{(m_1, n_1)}$ have B_j-intersection, $2 \leq j \leq k$. Let $A_s(L; G_1)$ be the set of $R \in [a_1 L, a_2 L]$ for which $\eta(R, G_1) = s$. For such R one can find $(m_1, n_1), \ldots, (m_s, n_s) \in \mathbb{Z}^2$ such that $R \in \bigcap_{j=1}^{s} \mathcal{D}_{(m_j, n_j)}$. The expectation (see Lemma 1)

$$E_L^{(k)} = E\eta^{(k)}(R, G_1) = \frac{1}{L(a_2 - a_1)} \sum_{s \geq k} C_s^k E\ell(A_s(G_1; L)) .$$

The crucial step is based upon the possibility of rewriting $E_L^{(k)}$ in a different way:

$$E_L^{(k)} = \frac{1}{L(a_2 - a_1)} \sum_{(m_1, n_1), \ldots, (m_k, n_k)} E\ell(\mathcal{D}_{(m_1, n_1)} \cap \ldots \cap \mathcal{D}_{(m_k, n_k)}) =$$

$$= \frac{1}{L(a_2 - a_1)} \sum_{(m_1, n_1)} \sum_{B_2, \ldots, B_k}^{(e)} \sum_{(m_2, n_2), \ldots, (m_k, n_k)} E\big[\ell(\mathcal{D}_{(m_1, n_1)} \cap \ldots \cap \mathcal{D}_{(m_\kappa, n_\kappa)}) \cdot$$

$$\cdot \chi_{(m_1, n_1), \ldots, (m_k, n_k)}(G_1; B_2, \ldots, B_k)\big] . \tag{7}$$

It is a technical part of the proof to show that one can restrict oneself by such $(m_1, n_1), \ldots, (m_k, n_k)$ that $\alpha(m_j, n_j) - \alpha(m_{j-1}, n_{j-1}) \geq \beta$ for some fixed β and after tending $L \to \infty$ let $\beta \to 0$ (see [3]). If we replace in the last expression $\ell(\mathcal{D}_{(m_1, n_1)} \cap \ldots \cap \mathcal{D}_{(m_\kappa, n_\kappa)})$ by $\ell(B_2, \ldots, B_k)$, the error has an absolute value not more than const $k\delta \cdot L^{-1}$. Take a partition of $[0, \frac{\pi}{2}]$ onto small intervals $\Delta^{(j)}$.

The main term in (7) takes the form

$$\widetilde{E}_L^{(k)} = \frac{1}{L(a_2 - a_1)} \sum_{m^{(1)}} \sum_{\Delta^{(j_1)} < \ldots < \Delta^{(j_\kappa)}}^{(\beta)} \sum_{B_2, \ldots, B_k}^{(e)} \cdot$$

$$\cdot \ell(B_2, \ldots, B_k) \cdot EN_{(m_1, n_1)}(\Delta^{(j_1)}, \ldots, \Delta^{(j_\kappa)}; B_2, \ldots, B_k)$$

where $\sum^{(\beta)}$ means that $\text{dist}(\Delta^{(j_{s_1})}, \Delta^{(j_{s_2})}) \geq \beta$. Using Lemmas 2 and 3 one easily derives that

$$\widetilde{E}_L^{(k)} = \frac{1}{n!} \int z_1^2 \ldots z_k^2 \pi_k(z_1, \ldots, z_\kappa \mid \alpha_1, \ldots, \alpha_\kappa) \Pi dz_j d\alpha_j + \gamma_2$$

where γ_2 tends to zero as $\beta \to 0$.

This gives theorem 1. Concerning the Poisson distribution for individual functions G_1 we have a weaker statement.

Theorem 2. *Under the conditions 1°-3° there exists a subsequence $\{L_j\}$, $L_j \to \infty$ as $j \to \infty$, and a set $\mathcal{G}_0 \subset \mathcal{G}$, $P(\mathcal{G}_0) = 1$ such that for any $G_1 \in \mathcal{G}_0$ the weak limit of distributions of $\eta(R, G_1)$, $a_1 L_j \leq R \leq a_2 L_j$ tends to the Poisson distribution with parameter $\frac{c}{2} \int G_1^2(\alpha) d\alpha$.*

Detailed proofs of Theorems 1 and 2 can be found in the paper [3].

§4 Quantum Kicked Rotator Model

One of the most popular models in quantum chaos is the model of the so-called quantum kicked rotator. It was introduced in the paper by Casati, Chirikov and Izraelev, Ford [4] as a quantum analog of the so-called Chirikov's or standard map. We shall describe now a slightly more general model and we claim that in a sense it is more natural. Write down the time-dependent Schrödinger equation

$$i\frac{\partial \psi(t, x)}{\partial t} = a\frac{\partial^2 \psi(t, x)}{\partial x^2} + ib\frac{\partial \psi(t, x)}{\partial x} + V(t, x)\psi(t, x) . \tag{8}$$

Here V is periodic in t and x, the period in time and x are taken to be equal to one. The function $\psi(t, x)$ is periodic in x with the same period 1. Most often people consider the case of $V(t, x) = \kappa \cos 2\pi x \sum_n \delta(t - n)$ which corresponds to a periodic sequence of kicks. The natural way to study the solutions of (8) is to consider the monodromy operator W, where $W\psi(t, x) = \psi(t + 1, x)$. It is a unitary operator acting in $\mathcal{L}^2(S^1)$ and for the case of the kicks $W = U_{a,b} \cdot W_{1,\kappa}$ where W_κ is the multiplication to $\exp\{i\kappa \cos 2\pi x\}$ while $U_{a,b} = e^{i(a\frac{d^2}{dx^2} + ib\frac{d}{dx})}$. After the Fourier transform $W_{1,\kappa}$ becomes a Toeplitz operator whose matrix elements decay very quickly outside the diagonal while $U_{a,b}$ is a diagonal operator with the matrix elements $\exp\left\{i(-4\pi^2 an^2 - 2\pi bn)\right\}\delta_{nm}$. It is convenient to change slightly the notations and consider a slightly more general case

$$W_x = U_{\alpha, z_1, z_2} W_{1,\kappa} \tag{9}$$

where $x = (x_1, x_2)$, $U_{\alpha, x_1, x_2} = \| \exp \{ 2\pi i (\alpha \frac{n(n-1)}{2} + nx_1 + x_2) \} \delta_{nm} \|$. Recall now that if $T_\alpha(x_1, x_2) = (x_1 + \alpha, x_2 + x_1)$ is the skew rotation of the two-dimensional torus then $T_\alpha^n(x, y) = (x_1 + n\alpha, x_2 + nx_1 + \frac{n(n-1)}{2}\alpha)$ and we see a close connection of (9) with the skew rotation.

We shall introduce now some general definitions which cover the case of (9) and the case of one-dimensional discrete Schrödinger operators with random potentials and which lead us to the notion of a random system. Start with a measure-preserving transformation T defined on a probability space $(M, \mathcal{M}, \mu)$, Assume that for each $x \in M$ we are given either a self-adjoint or a unitary operator B_x acting on the space ℓ_2 of sequences $f = \{f_n\}$. Denote by S the shift acting on ℓ_2, i.e. $(Sf)_n = f_{n+1}$.

Definition 3. The family $\{B_x\}$, $x \in M$ is connected with the dynamical system $(M, \mathcal{M}, \mu, T)$ if

$$SB_x = B_{Tx}S . \tag{10}$$

Let us give three examples which show the usefulness of the introduced notion.

1. The one-dimensional discrete Anderson model. Let $(M, \mathcal{M}, \mu)$ be the direct product of measure spaces, i.e. $x = \{x_n\}$ is a sequence of identically distributed random variables. Put

$$(B_x f)_n = -(f_{n+1} + f_{n-1}) + x_n \cdot f_n .$$

Definition 3 holds if T is the shift acting on M.

Thus M is infinite-dimensional. We have a rare situation where the infinite-dimensional case is simpler than a finite-dimensional one.

2. The one-dimensional discrete Schrödinger operator with a quasi-periodic potential. Here

$$(B_x f)_n = -(f_{n+1} + f_{n-1}) + V(n\omega + \alpha)f_n$$

and $V(\alpha)$ is a periodic function of period 1. Take $M = S^1$ and T to be a rotation to the angle ω. Again Definition 3 is applied.

3. In the two previous examples B_x were self-adjoint operators. Now take $M = \text{Tor}^2$ with the Lebesgue measure and $T = T_2$ being the skew rotation. Then the unitary operator W_x satisfies (10).

One of the first notions in the theory of random systems is the notion of the limiting density of states. It is well-known in the theory of Schrödinger operators with random potentials. For operators (9) it was constructed only recently by my student I. Koshovetz. The main idea was the following. Replace $W_{1,\kappa}$ by an operator which acts on the space of periodic sequences of the period N discretizing the Fourier transform. Let the result be denoted by $W_{1,\kappa}^{(N)}$. Then restrict also the diagonal operator U_{α,x_1,x_2} to the N-dimensional subspace and continue it periodically. If $U_{\alpha,x_1,x_2}^{(N)}$ is the corresponding finite-dimensional operator then the product $W_x^{(N)} = U_{\alpha,x_1,x_2}^{(N)} \cdot W_{1,\kappa}^{(N)}$ is the N-dimensional unitary operator and we may consider its spectrum. After that we can define the limiting density of state as the weak limit of the appearing finite-dimensional distributions. Koshovetz proves the existence of this limit as $N \to \infty$. I would be interesting to give a construction of the limiting density of states in the most general situation. The assumptions should include some assumptions concerning the dependence of B_x on x and properties of the approximation of T by periodic transformations.

Now we shall deal with the unitary operators W_x in (9) for $\kappa = 0$. It follows from the definitions that U_{α,x_1,x_2} is the diagonal operator with the diagonal matrix elements $\exp\left\{2\pi i\left(\frac{n(n-1)}{2}\alpha + nx_1 + x_2\right)\right\} = \lambda_n(\alpha, x_1, x_2)$. Following the general strategy we shall consider $-N \le n \le N$. Then we can introduce the characteristics of clustering of $\lambda_n(\alpha, x_1 x_2)$ in the same manner as in §1. In particular, take a constant c and introduce the set $A_k^{(N)}(\alpha, x_1, x_2)$ of $\lambda \in S^1$ such that $\Delta_\lambda = (\lambda, \lambda + \frac{c}{N})$ contains k numbers $\lambda_n(\alpha, x_1, x_2)$. We are interested in the behaviour of $\pi_k(N) = \int \ell\left(A_k^N(\alpha, x_1, x_2)\right) d\alpha\, dx_1\, dx_2$. Let $f(x) = \exp\{2\pi i x_2\}$, where $x = (x_1, x_2) \in \mathrm{Tor}^2$. Then $\lambda_n(\alpha, x_1 x_2) = f(T_\alpha^n x)$. Thus $\lambda \in A_k^{(N)}(\alpha, x_1, x_2)$ iff the inclusion $T_\alpha^n x \in \Delta_\lambda$ holds for k values of n among $-N \le n \le N$. Denote $p_k(N)$ the probability of this event assuming that we have the Lebesgue measure on the three-dimensional torus. By symmetry it does not depend on λ and $\pi_k(N) = p_k(N)$. Therefore we can consider $\lambda = 0$.

Now we shall show that the problem of study of the probability $p_k(N)$ is of the same nature as the problem which we discussed in §2, §3. Indeed, take the three-parameter family of functions $G(t; \alpha, x_1, x_2) = \frac{t(t-1)}{2}\alpha + tx_1 + x_2$. We are interested in the number of such n, $-N \le n \le N$, that

$\lambda_n(\alpha, x_1, x_2) \in (0, \frac{c}{N})$ or, in other words, in the number of such n that $m \le \frac{\alpha n(n-1)}{2} + nx_1 + x_2 \le m + \frac{c}{N}$ for some integer m. For every (α, x_1, x_2) consider the strip $\Pi_N(\alpha, x_1, x_2)$ bounded by the curves $G(t; \alpha, x_1 x_2)$ and $G(t; \alpha, x_1, x_2) + \frac{c}{N}$, $|t| \le -N$. Again this number is equal to the number of points of the two-dimensional lattice Z^2 belonging to $\Pi_N(\alpha, x_1, x_2)$. But now the strips $\Pi_N(\alpha, x_1, x_2)$ depend only on three independent random parameters α, x_1, x_2. This makes the analysis of the situation much harder.

It turns out that our problem is ultimately connected with some problem in number theory.

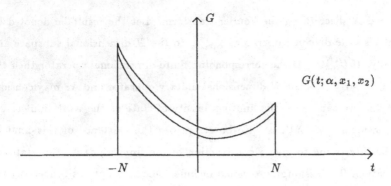

$$G(t; \alpha, x_1, x_2)$$

$$-N \qquad N \qquad t$$

Fig. 5

Consider the so-called double trigonometric sum of Weyl

$$S_N(a) = \frac{1}{N} \sum_{m=-N}^{N} \sum_{n=-N}^{N} \exp\{2\pi i m(a, n)\} \, ,$$

$a = (a_0, a_1, a_2)$, $(a, n) = a_0 n^2 + a_1 n + a_2$. It is easy to see that

$$\int S_N^2(a) da \leq \text{const} \, .$$

Therefore the sequence of probability distributions of $S_N(a)$ is weakly compact and one may pose the problem of finding the limiting probability distribution of $S_N(a)$ as $N \to \infty$. The summation over m can be done explicitly:

$$S_N(a) = \frac{1}{N} \sum_{n=-N}^{N} \frac{\exp\{2\pi i(N+1)(a, n)\} - \exp\{-2\pi i N(a, n)\}}{\exp\{2\pi i(a, n)\} - 1} =$$

$$= \frac{1}{N} \sum_{n=-N}^{N} \frac{\sin 2\pi N(a, n)}{tg\pi(a, N)} + \frac{1}{N} \sum_{n=-N}^{N} \cos 2\pi N(a, n) \, .$$

The last sum converges to zero at least in probability and we shall not consider it any more. The first sum can be rewritten as follows

$$S_N^{(1)}(a) = \frac{1}{N} \sum_{n=-N}^{N} \frac{\sin 2\pi N(a, n)}{tg\pi(a, n)} = \frac{1}{N} \sum_{n=-N}^{N} f_N(T_\alpha^n(x))$$

where $\alpha = 2a_0$, $x = (x_1, x_2)$ and $x_1 = a_1 + a_0$, $x_2 = a_2$, $f_N(x) = \frac{\sin 2\pi N x_2}{tg\pi x_2}$. Introduce

$$S_N^{(2)}(a) = \frac{1}{N} \sum_{\substack{n | |n| \leq N \\ \text{and } |(a,n)| \geq \frac{R}{N}}} f_N(T_\alpha^n x) \, .$$

Theorem 3. $\int \left|S_N^{(2)}(a)\right| da \leq \varepsilon_R$ *for all sufficiently large N where $\varepsilon_R \to 0$ as $R \to \infty$.*

Proof: Introduce the horizontal strip Π_0 of the width $2RN^{-1}$, $\Pi_0 = \{x \in \mathrm{Tor}^2 \mid |x_2| \leq RN^{-1}\}$ and the strips Π_k, $\Pi_k = \{x \in \mathrm{Tor}^2 \mid x_2 \in \Delta_k\}$ where, 1) $\Delta_{-k} = -\Delta_k$; 2) $\ell(\Delta_k) = \ell(\Delta_{-k}) = \frac{k}{N}$ except, perhaps, the last strip, $\Delta_1 < \Delta_2 < \ldots < \Delta_{k_0}$, $k_0 = \mathcal{O}(\sqrt{N})$. If θ_k is the left (right) boundary of Δ_k for $k > 0$ ($k < 0$) then

$$\theta_k = \frac{1}{N}\left(R + \frac{k(k-1)}{2}\right), \qquad k > 0,$$

$$\theta_k = -\frac{1}{N}\left(R + \frac{k(k-1)}{2}\right), \qquad k < 0.$$

Thus

$$S_N^{(2)}(a) = \sum_{k\mid 1 \leq |k| \leq k_0} \frac{1}{N} A_k, \qquad A_k = \sum_{n \mid T_\alpha^n(x) \in \Pi_k} f_N(T_\alpha^n x).$$

Denote also $h_N(x) = \sin 2\pi N x_2$. Now we can write

$$A_k = \frac{1}{tg\pi\theta_k}\sum_{n\mid T_\alpha^n x \in \Pi_k} h_N(T_\alpha^n x) + \sum_{n\mid T_\alpha^n x \in \Pi_k} h_N(T_\alpha^n x)\left(\frac{1}{tg\pi(a,n)} - \frac{1}{tg\pi\theta_k}\right) = A_k^{(1)} + A_k^{(2)}.$$

First we estimate $A_k^{(2)}$. We have

$$\left|ctg\pi(a,n) - ctg\pi\theta_k\right| \leq \frac{|(a,n) - \theta_k|}{\sin^2\pi\theta_k} \leq \frac{\ell(\Delta_k)}{\sin^2\pi\theta_k}.$$

Therefore

$$\left|\sum_k \frac{1}{N} A_k^{(2)}\right| \leq \frac{1}{N}\sum_k \frac{|\ell(\Delta_k)|}{\sin^2\pi\theta_k}\nu_\kappa(x)$$

where $\nu_\kappa(x) = \#\{n \mid T_\alpha^n x \in \Pi_k, |n| \leq N\}$. Thus $\int \nu_\kappa(x) da\, dx_1 dx_2 = (2N+1)$ mes $\Pi_k = \frac{2N+1}{N}k$ and

$$\int \left|\sum_k \frac{1}{N} A_k^{(2)}\right| da\, dx_1\, dx_2 \leq \mathrm{const} \sum \frac{k^2}{(R+k^2)^2} = \varepsilon_R^{(1)}.$$

Now come back to $A_k^{(1)} = \frac{1}{tg\pi\theta_k}\sum_{n\mid T_\alpha^n x \in \Pi_k} h_N(T_\alpha^n x)$. Decompose Π_k onto k^2 equal strips

$$\Pi_{k,j,\ell} = \left\{\frac{R}{N} + \frac{k(k-1)}{2N} + \frac{j}{N} + \frac{\ell}{kN} \leq x_2 < \frac{R}{N} + \frac{k(k-1)}{2} + \frac{j}{N} + \frac{\ell+1}{kN}\right\},$$

$0 \leq j < k$, $0 \leq \ell < k$, put $\Pi_{k,\ell}^{(1)} = \bigcup_{j=0}^{k-1} \Pi_{k,j,\ell}$ and write

$$A_k^{(1)} = \frac{1}{tg\pi\theta_k}\sum_{\ell=0}^{k-1} \sin 2\pi\frac{\ell}{k}\cdot\nu_{k,\ell}^{(1)}(x) + \frac{1}{tg\pi\theta_k}\sum_{\ell=0}^{k-1}\sum_{n\mid T_\alpha^n x \in \Pi_{k,\ell}^{(1)}} \left[h_N(T_\alpha^n x) - \sin 2\pi\frac{\ell}{k}\right] = A_k^{(3)} + A_k^{(4)}.$$

Here $\nu_{k,\ell}^{(1)}(x) = \#\{n \mid T_\alpha^n x \in \Pi_{k,\ell}^{(1)}, |n| \le N\}$. The sum $A_k^{(4)}$ is estimated in the same manner as above and we shall not reproduce the corresponding arguments in detail.

Consider $A_k^{(3)}$. Put

$$\chi_{\kappa,\ell}(x) = \begin{cases} 1 & x \in \Pi_{\kappa,\ell}^{(1)} \\ 0 & \text{otherwise}. \end{cases}$$

Then

$$\nu_{\kappa,\ell}^{(1)}(x) = \sum_{|n| \le N} \chi_{\kappa,\ell}(T_\alpha^n x) = \sum_{|n| \le N} \left[\chi_{\kappa,\ell}(T_\alpha^n x) - \text{mes}(\Pi_{k,\ell}^{(1)})\right] + (2N+1)\text{mes}\,\Pi_{k,\ell}^{(1)}.$$

This yields

$$A_k^{(3)} = \frac{1}{tg\pi\theta_\kappa} \sum_{\ell=0}^{k-1} \sin 2\pi \frac{\ell}{k} \cdot \text{mes}(\Pi_{k,\ell}^{(1)} + \frac{1}{tg\pi\theta_\kappa} \sum_{\ell=0}^{k-1} \sin 2\pi \frac{\ell}{k} \sum_{|n| \le N} \left[\chi_{k,\ell}(T_\alpha^n x) - \text{mes}(\Pi_{k,\ell}^{(1)})\right].$$

The measure of $\Pi_{k,\ell}^{(1)} = N^{-1}$ does not depend on ℓ. Therefore the first sum is equal to zero. Concerning the second sum we remark that it follows from the properties of the skew rotation T_α that the functions $\left[\chi_{\kappa,\ell}(T_\alpha^n x) - \text{mes}(\Pi_{k,\ell}^{(1)})\right]$ are mutually orthogonal for different n. Therefore

$$\|A_k^{(3)}\|_2^2 = \int |A_k^{(3)}|^2 \, d\alpha \, dx_1 dx_2 =$$

$$= \frac{1}{tg^2\pi\theta_k} \sum_{\ell=0}^{k-1} \left(\sin 2\pi \frac{\ell}{k}\right)^2 \cdot \int \sum_{\substack{|n_1| \le N \\ |n_2| \le N}} \left[\chi_{\kappa,\ell}(T_\alpha^{n_1} x) - \text{mes}(\Pi_{k,\ell}^{(1)})\right] \cdot$$

$$\cdot \left[\chi_{k,\ell}(T_\alpha^{n_2} x) - \text{mes}(\Pi_{k,\ell}^{(1)})\right] d\alpha \, dx_1 \, dx_2 \le \frac{\text{const}\, k}{tg^2\pi\theta_k}$$

and $\|A_k^{(3)}\|_2 \le \frac{\text{const}\sqrt{k}}{|tg\pi\theta_k|}$. Finally it gives

$$\left\| \sum_k \frac{1}{N} A_k^{(3)} \right\|_2 \le \text{const} \sum \frac{\sqrt{k}}{R+k^2} = \varepsilon_R^{(2)}.$$

which implies the statement of the theorem.

We shall use this theorem for the description of the limiting probability distributions for double trigonometric sums. Put $X_2 = x_2 N$ and fix R. Take such n_j that $x_2^{(j)} = \frac{n_j(n_j-1)}{2}\alpha + n_j x_1 + x_2 \in \left[-\frac{R}{N}, \frac{R}{N}\right] (\text{mod}\, 1)$. Then

$$S_N(a) = 2 \sum_j \frac{\sin 2\pi X_j}{2\pi X_j} + \gamma_R$$

where γ_R tends to zero in probability.

Hypothesis. The probability distribution of the random field $\{X_j\}$ converges to the Poisson field.

If this hypothesis is true then the limiting distribution for the double trigonometric sums is the distribution of the random variable $\zeta = 2\sum \frac{\sin 2\pi X_j}{2\pi X_j}$ where $X = \{X_j\}$ is a random realization of the Poisson field. The hypothesis also implies the exponential distribution of spacings between $\lambda_{\alpha,x_1,x_2}(n)$. I had a text with the proof of this hypothesis but P. Major found a gap in my calculations. I am sure that the results is true and have some ideas of how to correct the proof but at the moment it is better to consider the statement as an open question.

Final Remark. V.F. Lazutkin has recently informed me about his results concerning the quantum analog of KAM-Theory. They will be published in his forthcoming book. Previous results can be found in [5].

I thank the R. and B. Sackler foundation and Professor S. Abarbanel for the invitation to present these lectures. I also thank V. Milman for his efforts and constant help in arranging this visit.

References

1. M.V. Berry and M. Tabor. Proc. Roy. Soc. A 349 (1976), 101-123.
2. L.D. Landau, E.M. Lifschitz. Quantum Mechanics.
3. Ya.G. Sinai, Poisson Distribution in a Geometrical Problem. Advances in Soviet Mathematics, Publications of AMS (in press).
4. G. Casati, B.V. Chirikov, J. Ford, F.M. Izraelev, in: Stochastic Behaviour in Classical and Quantum Hamiltonian Systems, eds., G. Casati, J. Ford. Lecture Notes in Physics, 93, Springer-Verlag, p. 334-352.
5. V.F. Lazutkin, Convex Billiard and Eigenfunctions of the Laplace Operator, Publ. Leningrad University, Leningrad, 1981, 196pp. (in Russian).

QUASI-CLASSICAL EXPANSIONS
AND THE PROBLEM OF QUANTUM CHAOS

P.M. Bleher

Raymond and Beverly Sackler Faculty of Exact Sciences
Tel Aviv University, Tel Aviv, Israel
and
Dipartimento di Matematica Università di Roma "La Sapienza"

Abstract

In the present paper we comment some problems discussed in [1]. Following [1] we consider the distribution of eigenvalues E_{mn} of the Laplace-Beltrami operator on a two-dimensional revolution surface. We prove that the quasi-classical quantization rules give a correct asymptotic expansion for large E_{mn} and show that for the problem of quantum chaos two first terms of the quasi-classical expansion are essential. We specify a little bit the geometric problem studied in [1,5] to prove the Poisson distribution for the number ξ of the eigenvalues in the segment $[E, E + c]$, when $E \to \infty$, and show that the main theorem of [5] implies that for 'typical' revolution surfaces, $\xi = \xi^- + \xi^+$ where $\xi^- \equiv 0 \pmod 4$, $\xi^+ \equiv 0 \pmod 2$ and both $\frac{\xi^-}{4}$ and $\frac{\xi^+}{2}$ obey the Poisson distributions.

Contents

1.Introduction

The above-mentioned paper of Ya.G. Sinai [1] provides a general approach to the rigorous study of quantum chaos in the case when the underlying classical system is integrable. It contains many interesting results and beautiful problems and the aim of these notes is to specify and prove some of them. We begin with the eigenvalue problem (see [1])

$$v'' + (Ef^2 - n^2)v = 0 \qquad (1.1)$$

with the periodic boundary conditions

$$v(s + h) = v(s). \qquad (1.2)$$

Here $v = v(s)$, $s \in \mathbb{R}^1$, is an eigenfunction, $h > 0$ is a period, E is an eigenvalue, $f = f(s)$ is a given periodic, $f(s + h) = f(s)$, positive C^∞-smooth function on $\mathbb{R}^1$ and $n \in \mathbb{Z}$ is an integer external parameter of the problem. Because of the periodic boundary conditions we may consider Eq.(1.1) on the circle $S^1 = [0, h]$, $0 = h$. Now we shall present some results concerning the eigenvalue problem (1.1), (1.2).

Let $L^2(S^1; f^2(s)ds)$ be the Hilbert space of complex-valued functions $x(s)$ on $S^1 = [0, h]$, $0 = h$, with the scalar product

$$(x, y) = \int_0^h x(s)\overline{y(s)}f^2(s)ds.$$

Theorem 1. (see e.g. [2,3]). *There exists a countable set of real-valued C^∞-smooth eigenfunctions $v_m(s)$, $m = 0, 1, 2, ...$, of the problem (1.1), (1.2), which form an orthonormal basis in $L^2(S^1; f^2(s)ds)$. Corresponding eigenvalues E_m go to ∞ when $m \to \infty$.*

We shall assume that E_m are ordered, $0 \le E_0 \le E_1 \le ....$ Note that

$$E_0 \ge E_* \equiv \frac{n^2}{\max_{0 \le s \le h} f^2(s)}. \qquad (1.3)$$

Oscillating properties of the eigenfunctions $v_m(s)$ are described in the following theorem.

Theorem 2. *All zeroes of $v_m(s)$ are simple. The number of zeroes of $v_m(s)$ is even and it is equal to $2[(m + 1)/2]$.*

Remark that it is the 'periodic' version of a well-known result for the eigenvalue problem (1.1) with the Dirichlet boundary conditions (see e.g. [2,3]). With some alterations the proof given in [2] can be adapted to the periodic case (see Sect. 5 below).

2. Quasi-classical expansion

The quantum chaos concerns the distribution of eigenvalues $E_m = E_{m,n}$ when $E_{m,n} \to \infty$. Let

$$E = L^2 \varepsilon, \quad n = L\nu, \tag{2.1}$$

where L is a large parameter. Then (1.1) is read as

$$v'' + L^2(\varepsilon f^2 - \nu^2)v = 0. \tag{2.2}$$

Now we construct the quasi-classical (QC) expansion for v. To this end we write

$$v = A \exp(iL\Phi), \tag{2.3}$$

where $L\Phi$ is the phase and A is the amplitude, the both being real, and substitute it into (2.2). Since

$$v'' = (A'' + 2A'iL\Phi' + AiL\Phi'' - AL^2\Phi'^2) \exp(iL\Phi),$$

we get that

$$A'' - AL^2\Phi'^2 + L^2(\varepsilon f^2 - \nu^2)A + i(2A'\Phi' + A\Phi'')L = 0,$$

or

$$A'' + AL^2(-\Phi'^2 + \varepsilon f^2 - \nu^2) = 0, \tag{2.4}$$

$$2A'\Phi' + A\Phi'' = 0. \tag{2.5}$$

Solving the last equation we get (up to a constant multiplier)

$$A = \frac{1}{\sqrt{\Phi'}}. \tag{2.6}$$

To solve (2.4) we rewrite it as

$$\Phi'^2 = \varepsilon f^2 - \nu^2 + \frac{1}{L^2}\frac{A''}{A}. \tag{2.7}$$

Denote

$$\Psi = \Phi'. \tag{2.8}$$

Then $A = \frac{1}{\sqrt{\Psi}}$ so that

$$v(s) = \frac{1}{\sqrt{\Psi(s)}} \exp(iL \int_{s_0}^{s} \Psi(s')ds'). \tag{2.9}$$

After the substitution of $A = \frac{1}{\sqrt{\Psi}}$ into (2.7) we get that

$$\Psi^2 = \varepsilon f^2 - \nu^2 + \frac{1}{L^2}\left(-\frac{\Psi''}{2\Psi} + \frac{3}{4}\frac{\Psi'^2}{\Psi^2}\right). \tag{2.10}$$

Expand Ψ into the asymptotic series

$$\Psi = \Psi_0 + \frac{1}{L^2}\Psi_1 + \frac{1}{L^4}\Psi_2 + \dots$$

and equate in (2.10) the coefficients of the expansion in $\frac{1}{L^2}$-series. Then we get:

$$\Psi_0^2 = \varepsilon f^2 - \nu^2, \tag{2.11}$$

$$\Psi_1 = -\frac{\Psi_0''}{4\Psi_0^2} + \frac{3}{8}\frac{\Psi_0'^2}{\Psi_0^3} = -\left(\frac{\Psi_0'}{4\Psi_0^2}\right)' - \frac{1}{8}\frac{\Psi_0'^2}{\Psi_0^3}, \tag{2.12}$$

$$\dots$$

$$\Psi_k = \frac{1}{\Psi_0^3}Q_k(\Psi_0, \dots, \Psi_{k-1}, \Psi_0', \dots, \Psi_{k-1}', \Psi_0'', \dots, \Psi_{k-1}''), \tag{2.13}$$

$$\dots$$

where Q_k is a polynomial of $3k$ variables. To get (2.13) one can rewrite (2.10) as

$$\Psi^4 - (\varepsilon f^2 - \nu^2)\Psi^2 = \frac{1}{L^2}\left(-\frac{1}{2}\Psi\Psi'' + \frac{3}{4}\Psi'^2\right)$$

and equate the coefficients at $\frac{1}{L^{2k}}$. Then one obtains that

$$4\Psi_0^3\Psi_k - (\varepsilon f^2 - \nu^2)2\Psi_0\Psi_k = \hat{Q}_k, \tag{2.14}$$

where $\hat{Q}_k$ is a polynomial of $\Psi_0, \dots, \Psi_{k-1}''$. So by (2.11) one has that

$$2\Psi_0^3\Psi_k = \hat{Q}_k$$

or

$$\Psi_k = \frac{1}{\Psi_0^3}\frac{\hat{Q}_k}{2} \equiv \frac{1}{\Psi_0^3}Q_k, \tag{2.15}$$

what was stated.

Eqs. (2.11)-(2.13) determine Ψ_k recurrently. By (2.11)

$$\Psi_0 = \pm\sqrt{\varepsilon f^2 - \nu^2}. \tag{2.16}$$

Two cases are possible: (i) $\varepsilon f^2 - \nu^2 > 0$ for all $0 \le s \le h$; (ii) $\varepsilon f^2 - \nu^2 \le 0$ for some $0 \le s \le h$. The first case is simpler and we begin with it. Denote

$$\varepsilon^* = \frac{\nu^2}{\min_{0 \le s \le h} f^2(s)}. \tag{2.17}$$

Then the case (i) is characterized by the condition $\varepsilon > \varepsilon^*$, or

$$E > L^2 \varepsilon^*. \tag{2.18}$$

For definiteness we shall consider the sign $+$ in (2.16), i.e.

$$\Psi_0 = \sqrt{\varepsilon f^2 - \nu^2}. \tag{2.19}$$

Remark, that if $\varepsilon > \varepsilon^*$ then Ψ_0 is a smooth periodic function in s. Hence $\Psi_1, \Psi_2, \dots$ are also smooth periodic functions because of the recurrent equations (2.13). Consider the finite series

$$\Psi^{(k)} = \sum_{j=0}^{k} \frac{1}{L^{2j}} \Psi_j \tag{2.20}$$

and put

$$v^{(k)} = \frac{1}{\sqrt{\Psi^{(k)}}} \exp(iL \int \Psi^{(k)} ds). \tag{2.21}$$

$v^{(k)}$ is called the QC solution of Eq. (1.1) of the k-th order. If we substitute $\Psi^{(k)}$ into (2.10) we get the equation

$$\Psi^{(k)2} = \varepsilon f^2 - \nu^2 + \frac{1}{L^2}\left(-\frac{\Psi^{(k)\prime\prime}}{2\Psi^{(k)}} + \frac{3}{4}\frac{\Psi^{(k)\prime2}}{\Psi^{(k)2}}\right) + \frac{1}{L^{2k+2}}R_k \tag{2.22}$$

where

$$R_k = \frac{1}{\Psi^{(k)2}} P_k(\Psi_0, \dots, \Psi_k''; \frac{1}{L}), \tag{2.23}$$

where P_k is a polynomial of $3k + 4$ variables. It is worth to note for further use that $v^{(k)}$ satisfies the equation

$$v^{(k)\prime\prime} + L^2\left(\varepsilon f^2 - \nu^2 + \frac{1}{L^{2k+2}}R_k\right)v^{(k)} = 0. \tag{2.24}$$

The condition that $v^{(k)}$ is periodic in s is

$$L \int_0^h \Psi^{(k)}(s)ds = 2\pi m, \tag{2.25}$$

$m \in \mathbb{Z}$. For $k = 0$ it is reduced to

$$L \int_0^h \sqrt{\varepsilon f^2 - \nu^2}ds = \int_0^h \sqrt{E f^2 - n^2}ds = 2\pi m. \tag{2.26}$$

Eq. (2.25) is called the quantization condition of the k-th order.

Since the function

$$I(\varepsilon) \equiv \int_0^h \sqrt{\varepsilon f^2 - \nu^2}ds$$

is increasing in ε, Eq. (2.26) has a unique solution $\varepsilon = \varepsilon_m^{(0)}$ for any $m > m^* \equiv L\mu^*$, where

$$\mu^* = \frac{1}{2\pi} \int_0^h \sqrt{\varepsilon^* f^2 - \nu^2} ds. \tag{2.27}$$

$E_m^{(0)} = L^2 \varepsilon_m^{(0)}$ is called the quasi-eigenvalue of the zeroth order. Since

$$I(\varepsilon_m^{(0)}) = \frac{2\pi m}{L},$$

then

$$\varepsilon_{m+1}^{(0)} - \varepsilon_m^{(0)} = \frac{2\pi}{I'(\varepsilon_m^{(0)})} \frac{1}{L} + O\left(\frac{1}{L^2}\right),$$

so

$$E_{m+1}^{(0)} - E_m^{(0)} = \frac{2\pi}{I'(\varepsilon_m^{(0)})} L + O(1), \tag{2.28}$$

i.e. the distance between neighbor quazi-eigenvalues is of order of L.

Compute $E_m^{(1)}$. It is determined from the equation

$$\frac{2\pi}{m} L = \int_0^h \Psi^{(1)}(s) ds = \int_0^h \Psi_0(s) ds + \frac{1}{L^2} \int_0^h \Psi_1(s) ds.$$

Substituting formula (2.12) into this equation we get that

$$\frac{2\pi m}{L} = \int_0^h \sqrt{\varepsilon f^2 - \nu^2} ds - \frac{\varepsilon^2}{8L^2} \int_0^h \frac{(ff')^2}{(\varepsilon f^2 - \nu^2)^{5/2}} ds$$

$$= I(\varepsilon) - \frac{1}{L^2} G(\varepsilon), \tag{2.29}$$

where

$$G(\varepsilon) \equiv \frac{\varepsilon^2}{8} \int_0^h \frac{(ff')^2}{(\varepsilon f^2 - \nu^2)^{5/2}} ds. \tag{2.30}$$

Since $I'(\varepsilon) > 0$, the function $I(\varepsilon) - \frac{1}{L^2} G(\varepsilon)$ is increasing for large L, so Eq. (2.29) has a unique solution $\varepsilon = \varepsilon_m^{(1)}$ for large L. It can be written as

$$\varepsilon_m^{(1)} = \varepsilon_m^{(0)} + \frac{g(\varepsilon_m^{(0)})}{L^2} + O\left(\frac{1}{L^4}\right),$$

where

$$g(\varepsilon) = \frac{G(\varepsilon)}{I'(\varepsilon)} > 0. \tag{2.31}$$

Thus the quasi-eigenvalue of the first order is

$$E_m^{(1)} = L^2 \varepsilon_m^{(1)} = L^2 \varepsilon_m^{(0)} + g(\varepsilon_m^{(0)}) + O\left(\frac{1}{L^2}\right). \tag{2.32}$$

Similarly one can compute quasi-eigenvalues of higher order but subsequent corrections to $E_m^{(1)}$ are of order $O\left(\frac{1}{L^2}\right)$ and higher and they are not essential for the quantum chaos study.

The central problem now is: What is the correspondence between the eigenvalues E_m of the original problem (1.1), (1.2) and the quasi-eigenvalues $E_m^{(k)}$. Remark that the quasi-eigenvalues $E_m^{(k)}$ are twice degenerate. Namely, since Eq. (1.1) is real, both $v_m^{(k)}$ and

$$\overline{v_m^{(k)}} = \frac{1}{\sqrt{\Psi_m^{(k)}}} \exp(-iL \int \Psi_m^{(k)} ds), \quad \Psi_m^{(k)} \equiv \Psi^{(k)}\big|_{E=E_m^{(k)}},$$

are quasi-eigenfunctions. Now we can formulate the following main theorem.

Theorem 3. $\forall k \geq 0,\ 0.1 \geq \rho \geq 0,\ \lambda > 0,\ 0.1 \geq \sigma > 0,\ C > 0\ \exists M = M(k)$ and $\exists L_0 > 0$ such that $\forall L > L_0$ and $\forall E_m^{(0)}$ in the interval

$$(\varepsilon_* + \lambda L^{-\rho})L^2 \leq E_m^{(0)} \leq CL^2 \tag{2.33}$$

the following estimates hold:

$$|E_{2m-1} - E_m^{(k)}|, \quad |E_{2m} - E_m^{(k)}| \leq \frac{1}{L^{2k-\rho M(k)-\sigma}}.$$

Proof of this theorem will be given below in Sect. 5. Choosing $k = 1$ and $\delta = 1/2$, $\rho = \frac{1}{2M(1)}$, we get from Theorem 3 and Eq. (2.32) that for large L,

$$|E_{2m-1} - L^2\varepsilon_m^{(0)} - g(\varepsilon_m^{(0)})| \leq \frac{1}{L},$$

$$|E_{2m} - L^2\varepsilon_m^{(0)} - g(\varepsilon_m^{(0)})| \leq \frac{1}{L}. \tag{2.34}$$

3. Quasi-classical Expansion in the Presence of Turning Points

Consider now the case (ii),

$$L^2\varepsilon_* < E < L^2\varepsilon^*,$$

where

$$\varepsilon_* = \frac{\nu^2}{\max_{0 \leq s \leq h} f^2(s)}, \quad \varepsilon^* = \frac{\nu^2}{\min_{0 \leq s \leq h} f^2(s)}. \tag{3.1}$$

We shall assume that $f(s)$ has a unique point of maximum $s_{\max}$ and a unique point of minimum $s_{\min}$, the both being non-degenerate, i.e.

$$f''(s_{\max}) < 0, \quad f''(s_{\min}) > 0,$$

and $f'(s) \neq 0$ if $s \neq s_{\min}, s_{\max}$. Without loss of generality we may assume that $s_{\min} = 0$. Consider for any $\varepsilon_* < \varepsilon < \varepsilon^*$ the points $0 < a = a(\varepsilon) < b = b(\varepsilon) < h$ such that

$$\varepsilon f^2(a) - \nu^2 = \varepsilon f^2(b) - \nu^2 = 0. \tag{3.2}$$

a and b are called the turning points (see e.g. [4]). We have :

$$\varepsilon f^2(s) - \nu^2 > 0, \quad \text{if} \quad a < s < b,$$

$$\varepsilon f^2(s) - \nu^2 > 0, \quad \text{if} \quad s < a \quad \text{or} \quad s > b.$$

Let $\delta > 0$ be a small number. Consider the intervals

$$I_\delta(a) = S^1 \setminus [b - \delta, b + \delta],$$

$$I_\delta(b) = S^1 \setminus [a - \delta, a + \delta]$$

on the circle $S^1 = [0, h]$, $0 = h$. We construct QC solutions of Eq. (1.1) in $I_\delta(a)$ and in $I_\delta(b)$ and then we use matching relations for these solutions on their common part $I_\delta(a) \cap I_\delta(b)$ to obtain the quantization conditions of quasi-eigenvalues. Since

$$v'' + L^2(\varepsilon f^2 - \nu^2)v = 0 \tag{3.3}$$

and $\varepsilon f^2 - \nu^2$ has a simple zero at $s = a$, it is natural to construct the QC solution around $s = a$ with the help of the Airey function $\text{Ai}(x)$ which is the solution of the equation

$$\text{Ai}''(x) - x\,\text{Ai}(x) = 0,$$

such that

$$\lim_{x \to \infty} \text{Ai}(x) = 0.$$

Put

$$v(s) = \frac{1}{\sqrt{\varphi'(s)}} \text{Ai}(-L^{2/3}\varphi(s)). \tag{3.4}$$

Then Eq. (3.3) is reduced to

$$\varphi'^2 \varphi = \varepsilon f^2 - \nu^2 + \frac{1}{L^2}\left(-\frac{\varphi'''}{2\varphi'} + \frac{3}{4}\frac{\varphi''^2}{\varphi'^2}\right). \tag{3.5}$$

Expand φ into the asymptotic series in $\frac{1}{L^2}$,

$$\varphi = \varphi_0 + \frac{1}{L^2}\varphi_1 + \ldots$$

and equate in (3.5) the coefficients of the expansions in $\frac{1}{L^2}$-series. Then we get:

$$\varphi_0'^2 \varphi_0 = \varepsilon f^2 - \nu^2, \tag{3.6}$$

$$2\varphi_1' \varphi_0' \varphi_0 + \varphi_0'^2 \varphi_1 = -\frac{\varphi_0'''}{2\varphi_0'} + \frac{3}{4}\frac{\varphi_0''^2}{\varphi_0'^2}, \tag{3.7}$$

$$\cdots$$

$$2\varphi_k' \varphi_0' \varphi_0 + \varphi_0'^2 \varphi_k = \frac{1}{\varphi_0'^2} Q_k(\varphi_0, ..., \varphi_{k-1}'''), \tag{3.8}$$

$$\cdots$$

where Q_k is a polynomial of $4k$ variables. Now we can solve these equations recurrently and find uniquely smooth $\varphi_0, \varphi_1, ...$ Remark, that Eqs. (3.6)-(3.8) can be rewritten as

$$\left(\frac{3}{2}\varphi_0^{3/2\prime}\right)^2 = \varepsilon f^2 - \nu^2,$$

$$\left(\varphi_0^{1/2}\varphi_1\right)' = \frac{1}{2\varphi_0^{1/2}\varphi_0'}\left(-\frac{\varphi_0'''}{2\varphi_0'} + \frac{3}{4}\frac{\varphi_0''^2}{\varphi'^2}\right),$$

$$\cdots$$

$$\left(\varphi_0^{1/2}\varphi_k\right)' = \frac{1}{2\varphi_0^{1/2}\varphi_0'}\frac{1}{\varphi_0'^2} Q_k(\varphi_0, ..., \varphi_{k-1}'''),$$

$$\cdots$$

The smooth solutions of these equations are

$$\varphi_0(s) = \left|\frac{2}{3}\int_a^s \sqrt{|\varepsilon f^2 - \nu^2|}ds'\right|^{2/3} \text{sign}(s-a), \tag{3.9}$$

$$\varphi_1(s) = \frac{1}{|\varphi_0|^{1/2}}\int_a^s \frac{1}{2|\varphi_0|^{1/2}\varphi_0'}\left(-\frac{\varphi_0'''}{2\varphi_0'} + \frac{3}{4}\frac{\varphi_0''^2}{\varphi'^2}\right)ds', \tag{3.10}$$

$$\cdots$$

$$\varphi_k(s) = \frac{1}{|\varphi_0|^{1/2}}\int_a^s \frac{1}{2|\varphi_0|^{1/2}\varphi_0'}\frac{1}{\varphi_0'^2}Q_k(\varphi_0, ..., \varphi_{k-1}''')ds' \tag{3.11}$$

$$\cdots$$

Remark, that $\varphi_0, \varphi_1, ...$ are C^∞-smooth in $I_\delta(a)$ (including the point $s = a$!) and $\varphi_0'(a) \neq 0$.

Consider the finite series

$$\varphi^{(k)}(s) = \sum_{j=0}^k \frac{1}{L^{2j}}\varphi_j(s)$$

and put

$$v^{(k)}(s) = \frac{1}{\sqrt{\varphi^{(k)\prime}(s)}}\text{Ai}(-L^{2/3}\varphi^{(k)}(s)).$$

The function $v^{(k)}(s)$ is called the QC approximation of the k-th order for Eq. (1.1). Substituting $\varphi^{(k)}(s)$ into Eq. (3.5) we get that

$$\varphi^{(k)'2}\varphi^{(k)} = \varepsilon f^2 - \nu^2 + \frac{1}{L^2}\left(-\frac{\varphi^{(k)'''}}{2\varphi^{(k)'}} + \frac{3}{4}\frac{\varphi^{(k)''2}}{\varphi^{(k)'2}}\right) + \frac{1}{L^{2k+2}}R_k, \tag{3.12}$$

where

$$R_k = \frac{1}{\varphi^{(k)'2}}P_k(\varphi_0, ..., \varphi_k'''; \frac{1}{L}), \tag{3.13}$$

where P_k is a polynomial of $4k+4$ variables. Eq. (3.12) implies that $v^{(k)}$ satisfies the equation

$$v^{(k)''} + L^2(\varepsilon f^2 - \nu^2 + \frac{1}{L^{2k+2}}R_k)v^{(k)} = 0. \tag{3.14}$$

A similar QC expansion can be constructed in $I_\delta(b)$. A question arises how to match the expansions in $I_\delta(a)$ and $I_\delta(b)$. Consider a segment $[c, d]$ such that $a + \delta < c < d < b - \delta$, so that $[c, d] \subset I_\delta(a) \cap I_\delta(b) \cap [a, b]$. For large x we have the following QC asymptotics of $\mathrm{Ai}(-x)$:

$$\mathrm{Ai}(-x) = \frac{\mathrm{const}}{\sqrt{\zeta'(x)}}\cos(\zeta(x)),$$

where $\zeta(x)$ is the asymptotic series

$$\zeta(x) = -\frac{\pi}{4} + \frac{2}{3}x^{3/2} - \frac{5}{16}x^{-3/2} + ... = -\frac{\pi}{4} + \frac{2}{3}x^{3/2}\left(1 + \sum_{j=1}^{\infty}\alpha_j x^{-3j}\right). \tag{3.15}$$

It implies that for $s \in [c, d]$,

$$v(s) = \frac{\mathrm{const}}{\sqrt{\varphi'(s)\zeta'(x)}}\cos(\zeta(L^{2/3}\varphi(s))) = \frac{\mathrm{const}}{\sqrt{\Phi'(s)}}\cos(L\Phi(s)), \tag{3.16}$$

where

$$\Phi(s) = L^{-1}\zeta(L^{2/3}\varphi(s)) = -\frac{\pi}{4L} + \frac{2}{3}\varphi^{3/2} - \frac{5}{16L^2}\varphi^{-3/2} + ... =$$

$$= -\frac{\pi}{4L} + \frac{2}{3}(\varphi_0 + \frac{1}{L^2}\varphi_1 + ...)^{3/2} - \frac{5}{16L^2}(\varphi_0 + \frac{1}{L^2}\varphi_1 + ...)^{-3/2} + ...$$

$$= -\frac{\pi}{4L} + \frac{2}{3}\varphi_0^{3/2} + \frac{1}{L^2}(\varphi_0^{1/2}\varphi_1 - \frac{5}{16}\varphi_0^{-3/2} +) + ...$$

Substituting the expressions (3.9), (3.11) for $\varphi_0, \varphi_1, ...$ into the last formula, we get that

$$\Phi(s) = -\frac{\pi}{4L} + \int_a^s \sqrt{\varepsilon f^2 - \nu^2}\,ds'$$

$$+ \frac{1}{L^2}\left[\int_a^s \frac{1}{\varphi_0^{1/2}}\left(-\frac{1}{4}\frac{\varphi_0'''}{\varphi_0'^2} + \frac{3}{8}\frac{\varphi_0''2}{\varphi_0'^3}\right)ds' - \frac{5}{72}\left(\int_a^s \sqrt{\varepsilon f^2 - \nu^2}\,ds'\right)^{-1}\right] + ...$$

$$= -\frac{\pi}{4L} + \Phi_0(s) + \frac{1}{L^2}\Phi_1(s) + \frac{1}{L^4}\Phi_2(s) + ... \tag{3.17}$$

Put

$$\Psi(s) \equiv \Phi'(s) = \Phi'_0(s) + \frac{1}{L^2}\Phi'_1(s) + \frac{1}{L^4}\Phi'_2(s) + \ldots$$

$$= \Psi_0(s) + \frac{1}{L^2}\Psi_1(s) + \frac{1}{L^4}\Psi_2(s) + \ldots . \tag{3.18}$$

Remark, that for $s \in [c,d]$, $\quad \varepsilon f^2 - \nu^2 > 0$, hence (3.16), (3.17) give a QC expansion of the form (2.9). Hence the coefficients $\Psi_0, \Psi_1, \ldots$ are defined by Eqs. (2.11), (2.13). In particular,

$$\Psi_0 = p \equiv \sqrt{\varepsilon f^2 - \nu^2},$$

$$\Psi_1 = -\left(\frac{p'}{4p^2}\right)' - \frac{1}{8}\frac{p'^2}{p^3}. \tag{3.19}$$

Remark, that Ψ_1 has a non-integrable singularity at $s = a$, $\Psi_1(s) \sim \frac{\text{const}}{(s-a)^{5/2}}$ because of $p \sim \text{const}(s-a)^{1/2}$. An analysis of (3.17) shows that

$$\Phi_1(s) = -\frac{p'(s)}{4p^2(s)} - \text{reg}\int_a^s \frac{p'^2}{8p^3}ds', \tag{3.20}$$

where the regularization of the last integral is understood as

$$\text{reg}\int_a^s \frac{p'^2}{p^3}ds' = \lim_{\delta\to+0}\left(\int_{a+\delta}^s \frac{p'^2}{p^3}ds' - \frac{C_0}{\delta^{3/2}} - \frac{C_1}{\delta^{1/2}}\right),$$

where the constants C_0, C_1 are chosen in such a way that the limit does exist. Formulae (3.16), (3.17) were obtained starting from the turning point $s = a$. Similar formulae can be obtained starting from $s = b$:

$$v(s) = \frac{\text{const}}{\sqrt{\Omega'(s)}}\cos(L\Omega(s)), \tag{3.21}$$

where

$$\Omega(s) = -\frac{\pi}{4L} + \int_s^b \sqrt{\varepsilon f^2 - \nu^2}\,ds'$$

$$+ \frac{1}{L^2}\left[\int_s^b \frac{1}{\varphi_0^{1/2}}\left(-\frac{1}{4}\frac{\varphi_0'''}{\varphi_0'^2} + \frac{3}{8}\frac{\varphi_0''^2}{\varphi_0'^3}\right)ds' - \frac{5}{72}\left(\int_s^b \sqrt{\varepsilon f^2 - \nu^2}\,ds'\right)^{-1}\right] + \ldots$$

$$= -\frac{\pi}{4L} + \Omega_0(s) + \frac{1}{L^2}\Omega_1(s) + \frac{1}{L^4}\Omega_2(s) + \ldots . \tag{3.22}$$

Moreover, similarly to (3.16), (3.20) we have that

$$\Omega'(s) = -\Psi = -\sqrt{\varepsilon f^2 - \nu^2} - \frac{1}{L^2}\Psi_1 - \frac{1}{L^4}\Psi_2 - \ldots, \tag{3.23}$$

and

$$\Omega_1(s) = \frac{p'(s)}{4p^2(s)} - \text{reg}\int_s^b \frac{p'^2}{8p^3}ds', \tag{3.24}$$

where the regularization is made at $s = b$.

Since by (3.19), (3.23)

$$\Phi'(s) = -\Omega'(s) = \Psi(s),$$

then

$$\Phi(s) + \Omega(s) = \text{const}.$$

To match expansions (3.16) and (3.21) we have to choose here const $= \frac{\pi m}{L}$, i.e.

$$\Phi(s) + \Psi(s) = \frac{\pi m}{L}, \tag{3.25}$$

$m = 0, 1, \dots$. This is the quantization condition, which defines the QC eigenvalue. Substituting expansions (3.17), (3.22) of $\Phi(s)$, $\Psi(s)$ we get that

$$-\frac{\pi}{2L} + \int_a^b pds - \frac{1}{L^2}\text{reg}\int_a^b \frac{p'^2}{8p^3}ds + \dots = \frac{\pi m}{L},$$

or

$$\int_a^b pds - \frac{1}{L^2}\text{reg}\int_a^b \frac{p'^2}{8p^3}ds + \dots = \frac{\pi\left(m+\frac{1}{2}\right)}{L}, \tag{3.26}$$

where $p = \sqrt{\varepsilon f^2 - \nu^2}$. Put

$$\varepsilon_m = \varepsilon_m^{(0)} + \frac{1}{L^2}\varepsilon_{m1} + \frac{1}{L^4}\varepsilon_{m2} + \dots$$

and substitute this asymptotic expansion into the last equation. Then we can find recurrently the coefficients $\varepsilon_m^{(0)}$, ε_{mk}, $k \geq 1$. The zero coefficient $\varepsilon = \varepsilon_m^{(0)}$ is the solution of the equation

$$L\int_a^b \sqrt{\varepsilon f^2 - \nu^2}ds = \pi\left(m + \frac{1}{2}\right),$$

which is the famous Bohr-Sommerfeld quantization condition. It can be written also as

$$\int_a^b \sqrt{Ef^2 - n^2}ds = \pi\left(m + \frac{1}{2}\right). \tag{3.27}$$

Next,

$$\varepsilon_{m1} = g(\varepsilon_m^{(0)}), \tag{3.28}$$

where

$$g(\varepsilon) = \frac{G(\varepsilon)}{F'(\varepsilon)},$$

$$F(\varepsilon) = \int_a^b p\,ds = \int_a^b \sqrt{\varepsilon f^2 - \nu^2}\,ds,$$

$$G(\varepsilon) = \text{reg}\int_a^b \frac{p'^2}{8p^3}\,ds.$$

Similarly one can compute $\varepsilon_{m2}, \varepsilon_{m3}, \dots$.

Now we can formulate the following main theorem.

Theorem 4. $\forall k \geq 0$, $0.1 \geq \rho \geq 0$, $\lambda > 0$, $0.1 > \sigma > 0$, $\exists M = M(k)$ and $\exists L_0 > 0$ such that $\forall L > L_0$ and $\forall E_m^{(0)}$ in the interval

$$L^2(\varepsilon_* + \lambda L^{-\rho}) < E_m^{(0)} < L^2(\varepsilon^* - \lambda L^{-\rho}), \tag{3.29}$$

which satisfies the Bohr-Sommerfeld quantization condition (3.27), the following estimate holds:

$$|E_m - E_m^{(k)}| < \frac{1}{L^{2k-\rho M(k)-\sigma}}.$$

Choosing $k = 1$ and $\sigma = 0.1$, $\rho = \frac{1}{2M(1)}$, we get from Th. 4 that for large L,

$$|E_m - L^2\varepsilon_m^{(0)} - g(\varepsilon_m^{(0)})| \leq \frac{1}{L}. \tag{3.30}$$

Proof of this theorem will be given in the next Section.

4. Proofs

Proof of Theorem 2.   In [2] a similar theorem is proved for Dirichlet's boundary conditions. We show now how to adapt the proof given in [2] to periodic boundary conditions.

1. All zeroes of $v_m(s)$ are simple. It follows from the uniqueness theorem for the Cauchy problem for the Eq. (1.1).

2. The number of zeroes is even. Since zeroes are simple, each zero is the change of sign. By the periodicity the number of the sign changes is even, so the number of zeroes is the same.

3. The number of zeroes does not exceed $2\left[\frac{m+1}{2}\right]$. It is proved in [2] that the number of intervals between neighbor zeroes of the eigenfunction v_m does not exceed $m + 1$. Since it is even, it does not exceed $2\left[\frac{m+1}{2}\right]$, so the number of zeroes of v_m does not exceed $2\left[\frac{m+1}{2}\right]$.

4. The number of zeroes is not less than $2\left[\frac{m+1}{2}\right]$. In [2], pp. 463–464, it is proved that the number of zeroes of v_m is not less than m. Since it is even, it is not less than $2\left[\frac{m+1}{2}\right]$. Theorem 2 is proved.

Proof of Theorem 3.   Proof consists of 'spectral' and 'differential' parts. In the 'spectral' part we shall prove that near any quasi-eigenvalue $E_m^{(k)}$ at least two genuine eigenvalues lie. In the 'differential' part we shall prove on the contrary that if E is an eigenvalue which lies near $E_m^{(0)}$ then the corresponding eigenfunction has $2m$ zeroes. It enables us to prove Theorem 3.

'Spectral' part. Let

$$v_m^{(k)} = \frac{1}{\Psi_m^{(k)}} \exp\left(iL \int \Psi_m^{(k)} ds\right) \tag{4.1}$$

be the periodic quasi-eigenfunction of the k-th order with the quasi-eigenvalue $E_m^{(k)}$. Then by (2.24)

$$-v_m^{(k)''} + (n^2 - E_m^{(k)}f^2)v_m^{(k)} = \frac{1}{L^{2k}}R_k v_m^{(k)}. \tag{4.2}$$

Let us show that R_k satisfies the estimate

$$|R_k| < \text{const } L^{\rho M(k)}. \tag{4.3}$$

To this end let us first prove that

$$|\Psi_k| < \text{const } L^{\rho M_0(k)}. \tag{4.4}$$

Let $\varepsilon = \varepsilon_m^{(k)}$. From (2.33) it follows that

$$\Psi_0 = \sqrt{\varepsilon f^2 - \nu^2} > \text{const } L^{-1.1\rho}.$$

It implies that

$$|\frac{d^j \Psi_0}{ds^j}| < \text{const } L^{M_1(j)\rho}.$$

Now the estimates (4.4), (4.3) are proved easily by induction in k. We omit the details.

Estimate (4.3) implies that

$$\| -v_m^{(k)''} + (n^2 - E_m^{(k)}f^2)v_m^{(k)} \|_{L^2([0,h])} \leq \frac{1}{L^{2k}} \| R_k v_m^{(k)} \|_{L^2([0,h])}$$
$$\leq \frac{\text{const}}{L^{2k-\rho M(k)}} \| f v_m^{(k)} \|_{L^2([0,h])}. \tag{4.5}$$

This inequality can be interpreted in the following way. Consider the differential operator

$$T: v \to \frac{1}{f^2}(-v'' + n^2 v),$$

which is self-adjoint in $L^2([0,h]; f^2(s)ds)$. Then (4.5) is equivalent to

$$\| T v_m^{(k)} - E_m^{(k)} v_m^{(k)} \| \leq \frac{\text{const}}{L^{2k-\rho M(k)}} \| v_m^{(k)} \|, \tag{4.6}$$

where $\| \cdot \| = \| \cdot \|_{L^2([0,h]; f^2(s)ds)}$. Besides, we have the general spectral inequality

$$\| T v - E v \| \geq \| v \| \cdot \text{dist}(E, \text{Spec} T). \tag{4.7}$$

Comparing it with (4.6) we get that

$$\text{dist}(E_m^{(k)}, \{E_j, j = 0, 1, 2, ...\}) \leq \frac{\text{const}}{L^{2k-\rho M(k)}},$$

or

$$|E_m^{(k)} - E_j| \le \frac{\text{const}}{L^{2k-\rho M(k)}}$$

for some j.

Let us show that there is also another eigenvalue but E_j which lies near $E_m^{(k)}$. We use here that $E_m^{(k)}$ is twice degenerate. Let

$$v_m^{(k)} = \sum_{i=1}^{\infty} a_i v_i = a_j v_j + v'.$$

Since $v_j(s)$ are real-valued, then

$$\overline{v_m^{(k)}} = \sum_{i=1}^{\infty} \overline{a_i} v_i = \overline{a_j} v_j + \overline{v'}.$$

Differentiating by parts we get that

$$(v_m^{(k)}, \overline{v_m^{(k)}}) \equiv \int_0^h \exp(2iL \int \Psi_m^{(k)} ds') \frac{f^2(s)ds}{\Psi_m^{(k)}} = O(\frac{1}{L^N}) \tag{4.8}$$

for any $N \ge 0$, so, in particular, for $N = 1$ we have that

$$(v_m^{(k)}, \overline{v_m^{(k)}}) = a_j^2 + (v', \overline{v'}) = O(\frac{1}{L}).$$

Since

$$|a_j|^2 + \| v' \|^2 = \| v_m^{(k)} \|^2 = \int_0^h \frac{f^2}{\Psi_m^{(k)}} ds$$

is of order of 1, it implies that for large L

$$\| v_m^{(k)} \| \le 2 \| v' \|.$$

Now, by the general spectral inequality

$$\| (T - E_j)v' \| \ge \| v' \| \cdot \text{dist}(E_j, \{E_i, i \ne j\}).$$

Since

$$\| (T - E_j)v' \| = \| (T - E_j)v_m^{(k)} \| \le \| (T - E_m^{(k)})v_m^{(k)} \| + |E_m^{(k)} - E_j| \| v_m^{(k)} \|$$

$$\le \frac{\text{const}}{L^{2k-\rho M(k)}} \| v_m^{(k)} \|,$$

it implies that

$$\text{dist}(E_j, \{E_i, i \ne j\}) \le \frac{\text{const}}{L^{2k}} \cdot \frac{\| v_m^{(k)} \|}{\| v' \|} \le 2\frac{\text{const}}{L^{2k-\rho M(k)}}.$$

Hence there exists an eigenvalue $E_i, i \neq j$, such that

$$|E_i - E_m^{(k)}| \leq \frac{\text{const}}{L^{2k-\rho M(k)}}.$$

Thus we have shown that in $O\left(\frac{1}{L^{2k-\rho M(k)}}\right)$-neighborhood of $E_m^{(k)}$ at least two eigenvalues E_j, E_i lie.

'Differential part'. Consider now an arbitrary eigenvalue E in the interval

$$(\varepsilon_* + \lambda L^{-\rho}) \leq E \leq CL^2.$$

Assume that the corresponding eigenvalue has $2m$ zeroes. We shall prove that in such a case E is close to $E_m^{(k)}$. For simplicity of formulae we shall consider the case $\rho = 0$. The general case is treated in the same manner. Let

$$E_\pm = E \pm \frac{1}{L^{2k-\sigma}} \tag{4.9}$$

and $\Psi_\pm(s) = \Psi^{(k)}(s; E_\pm)$ be the QC phase function of the k-th order at $E = E_\pm$. Assume that s_0 is a zero of $v(s)$. Consider the QC real eigenfunctions

$$v_\pm(s) = \frac{1}{\sqrt{\Psi_\pm(s)}} \sin(L \int_{s_0}^s \Psi_\pm(s')ds'). \tag{4.10}$$

Remark that

$$v_\pm(s_0) = v(s_0) = 0. \tag{4.11}$$

We shall show now that any zero of $v(s)$ lies between the corresponding zeroes of $v_-(s)$ and $v_+(s)$.

Denote

$$q = Ef^2 - n^2,$$
$$q_\pm = E_\pm f^2 - n^2 + \frac{1}{L^{2k}}R_{k,\pm}, \tag{4.12}$$

so that

$$v'' + qv = 0,$$
$$v_\pm'' + q_\pm v_\pm = 0. \tag{4.13}$$

In virtue of (4.3) and (4.9),

$$0 < q_- < q < q_+ \tag{4.14}$$

for large L. Put

$$\frac{v}{v'} = \tan \phi, \quad \frac{v_\pm}{v'_\pm} = \tan \phi_\pm.$$

Then (4.13) implies that

$$\phi' = q \sin^2 \phi + \cos^2 \phi,$$

$$\phi'_\pm = q_\pm \sin^2 \phi_\pm + \cos^2 \phi_\pm. \tag{4.15}$$

By (4.11)

$$\phi(s_0) = \phi_\pm(s_0) = 0. \tag{4.16}$$

Since for any ϕ,

$$q_- \sin^2 \phi + \cos^2 \phi \leq q \sin^2 \phi + \cos^2 \phi \leq q_+ \sin^2 \phi + \cos^2 \phi$$

we get from (4.15), (4.16) that

$$\phi_-(s) \leq \phi(s) \leq \phi_+(s) \tag{4.17}$$

for any $s > s_0$.

Since $q > 0$, $q_\pm > 0$, (4.15) implies that $\phi(s)$, $\phi_\pm(s)$ are increasing. The l-th zeroes of $\phi(s)$ and of $\phi_\pm(s)$ are the solutions of the equations

$$\phi(s_l) = \pi l,$$

$$\phi_\pm(s_{l,\pm}) = \pi l,$$

respectively. Thus we get from (4.17) that for $l \geq 0$,

$$s_{l,-} \leq s_l \leq s_{l,+}.$$

We assume that v has $2m$ zeroes, therefore $s_{2m} = s_0 + h$, so

$$s_{2m,-} \leq s_0 + h \leq s_{2m,+}.$$

By (4.10)

$$L \int_{s_0}^{s_{2m,\pm}} \Psi^{(k)}(s; E_\pm) ds = 2\pi m,$$

so the last inequality implies that

$$L \int_{s_0}^{s_0+h} \Psi^{(k)}(s; E_-) ds \leq 2\pi m \leq L \int_{s_0}^{s_0+h} \Psi^{(k)}(s; E_+) ds.$$

Since $E_m^{(k)}$ is defined by the equation

$$L \int_{s_0}^{s_0+h} \Psi^{(k)}(s; E_m^{(k)}) ds = 2\pi m$$

and $\int_{s_0}^{s_0+h} \Psi^{(k)}(s; E) ds$ is increasing with E, we get that $E_- \leq E_m^{(k)} \leq E_+$, or

$$|E_m^{(k)} - E| \leq \frac{1}{L^{2k-\sigma}}. \tag{4.18}$$

Thus we have prove that if E lies in the interval

$$(\varepsilon_* + \lambda)L^2 \leq E \leq CL^2. \tag{4.19}$$

and v has $2m$ zeroes then (4.18) holds.

Summing up the both parts of the proof we get that for large L:

(i) In $\frac{1}{L^{2k-\sigma}}$-neighborhood of any $E_m^{(k)}$ in the interval (4.19) at least two eigenvalues E_j, E_i lie;

(ii) If E_j lies in the interval (4.19) and v_j has $2m$ zeroes then $|E_j - E_m^{(k)}| \frac{1}{L^{2k-\sigma}}$.

Since by Theorem 2 v_{2m-1}, v_{2m} have $2m$ zeroes and by (2.28) $E_{m+1} - E_m > \text{const } L$, (i), (ii) imply that

$$|E_{2m-1} - E_m^{(k)}|, \quad |E_{2m} - E_m^{(k)}| < \frac{1}{L^{2k-\sigma}}.$$

This proves Theorem 3.

Proof of Theorem 4. For simplicity we shall consider the case $\rho = 0$. The general case is considered in the same manner with minor modifications. Let

$$\mu_*(\lambda) = \int_{a(\varepsilon_*(\lambda))}^{b(\varepsilon_*(\lambda))} \sqrt{\varepsilon_*(\lambda)f^2 - \nu^2} ds,$$

$$\mu^*(\lambda) = \int_{a(\varepsilon^*(\lambda))}^{b(\varepsilon^*(\lambda))} \sqrt{\varepsilon^* f^2 - \nu^2} ds,$$

where $\varepsilon_*(\lambda) = \varepsilon_* + \lambda$, $\varepsilon^*(\lambda) = \varepsilon^* - \lambda$. Put

$$m_*(\lambda) = L\mu_*(\lambda) - \frac{1}{2}, \quad m^*(\lambda) = L\mu^*(\lambda) - \frac{1}{2}.$$

Then the condition (3.29) is equivalent to

$$m_*(\lambda) \leq m \leq m^*(\lambda). \tag{4.20}$$

Consider QC approximations in the segment $I_\delta(a) = S^1 \setminus [b - \delta, b + \delta]$,

$$v^{(k)} = \frac{1}{\sqrt{\varphi^{(k)\prime}}} \mathrm{Ai}(-L^{2/3} \varphi^{(k)}), \tag{4.21}$$

$$\varphi^{(k)} = \sum_{j=0}^{k} \frac{1}{L^{2j}} \varphi_j,$$

and in the segment $I_\delta(b) = S^1 \setminus [a - \delta, a + \delta]$,

$$w^{(k)} = \frac{1}{\sqrt{\omega^{(k)\prime}}} \mathrm{Ai}(-L^{2/3} \omega^{(k)}), \tag{4.22}$$

$$\omega^{(k)} = \sum_{j=0}^{k} \frac{1}{L^{2j}} \omega_j,$$

and try to arrange from them a unified function on S^1. Let

$$m_1 = \left[\frac{m}{2}\right], \quad m_2 = m - m_1 + 1. \tag{4.23}$$

We state that ε exists in the interval

$$|\varepsilon - \varepsilon_m^{(k)}| < \frac{1}{L^{2k+2-\sigma}} \tag{4.24}$$

such that the m_1-th zero of $v^{(k)}$ (counting out from $s = a$) coincides with the m_2-th zero of $w^{(k)}$ (counting out from $s = b$).

The asymptotic expansion of the m_1-th zero s_{m_1} of $v^{(k)}$ in $\frac{1}{L^2}$-series,

$$s_{m_1} = s_{m_1,0} + \frac{1}{L^2} s_{m_1,1} + \frac{1}{L^4} s_{m_1,2} + \cdots,$$

can be found from the equation

$$L\Phi(s) \equiv \zeta(L^{2/3}\varphi(s)) = -\frac{\pi}{2} + m_1\pi. \tag{4.25}$$

Similarly, the m_2-th zero of $w^{(k)}$ can be found from the equation

$$-L\Omega(s) \equiv \zeta(-L^{2/3}\omega(s)) = -\frac{\pi}{2} + m_2\pi. \tag{4.26}$$

For

$$\varepsilon_m = \varepsilon_m^{(0)} + \sum_{j=1}^{\infty} L^{-2j} \varepsilon_{mj},$$

which is determined by the quantization condition (3.25), the equality (4.25) implies the one (4.26), so in that case the m_1-th zero of v coincides with the m_2-th zero of w. It implies that if $s_{m_1}^{(k)}$ is the m_1-th zero of $v^{(k)}(s; \varepsilon_m^{(k)})$, and $\tilde{s}_{m_2}^{(k)}$ is the m_2-th zero of $w^{(k)}(s; \varepsilon_m^{(k)})$, where

$$\varepsilon_m^{(k)} = \varepsilon_m^{(0)} + \sum_{j=1}^{k} L^{-2j}\varepsilon_{mj},$$

then

$$|s_{m_1}^{(k)} - \tilde{s}_{m_2}^{(k)}| < CL^{2h-2}. \tag{4.27}$$

Next, $s = s_{m_1}^{(0)}$ is the solution of the equation

$$L \int_a^s \sqrt{\varepsilon f^2 - v^2}\, ds' = \pi\left(m + \frac{1}{2}\right)$$

and it is easy to see that

$$\frac{ds_{m_1}^{(0)}}{d\varepsilon} > C > 0,$$

so for large L

$$\frac{ds_{m_1}^{(k)}}{d\varepsilon} = \frac{ds_{m_1}^{(0)}}{d\varepsilon} + O\left(\frac{1}{L^2}\right) > C_0 > 0.$$

Similarly,

$$\frac{d\tilde{s}_{m_2}^{(k)}}{d\varepsilon} < -C_0 < 0.$$

Therefore because of (4.27), ε exists in the interval (4.24) such that $s_{m_1}^{(k)} = \tilde{s}_{m_2}^{(k)}$, which was stated. Denote this ε by $\hat{\varepsilon}_m^{(k)}$ and $s_{m_1}^{(k)} = \tilde{s}_{m_2}^{(k)}$ by $\hat{s}_m^{(k)}$. Let $\hat{E}_m^{(k)} = L^2\hat{\varepsilon}_m^{(k)}$. By (4.24)

$$|\hat{E}_m^{(k)} - E_m^{(k)}| < \frac{1}{L^{2k-\sigma}}. \tag{4.28}$$

Let $\chi(s) \in C^\infty(S^1)$ be such a function that $\chi(s) \equiv 0$ in a neighborhood of $s = 0$ and $\chi(s) \equiv 1$ in a neighborhood of the segment $[a, b]$, where $a = a(\hat{\varepsilon}_m^{(k)})$, $b = b(\hat{\varepsilon}_m^{(k)})$. Put

$$\hat{v}_m^{(k)}(s) = \begin{cases} v_m^{(k)}(s)\chi(s) & \text{if } 0 \le s \le \hat{s}_m^{(k)}, \\ C_m^{(k)} w_m^{(k)}(s)\chi(s) & \text{if } \hat{s}_m^{(k)} \le s \le h, \end{cases} \tag{4.29}$$

where the constant $C_m^{(k)}$ is taken in such a way that $\frac{d\hat{v}_m^{(k)}}{ds}(s)$ has no jump at $s = \hat{s}_m^{(k)}$. It is easy to see that $C_m^{(k)} = (-1)^m + O(L^{-2k})$. We view $v_m^{(k)}(s)$ as a unique QC approximation of Eq. (1.1) on S^1. Remark, that $\frac{d^2\hat{v}_m^{(k)}}{ds^2}$ can have a jump at $s = \hat{s}_m^{(k)}$ but it does not essential in what follows.

'Spectral' part of the proof. Let c, d be such that $0 < c < a < b < d < h$ and $\chi(c) = \chi(d) = 1$. Consider three regions:

$$D_1 = \{c \leq s \leq s_m^{(k)}\},$$
$$D_2 = \{s_m^{(k)} \leq s \leq d\},$$
$$D_3 = \{J0 \leq s \leq c \quad \text{or} \quad d \leq s \leq h\}.$$

In D_1 $\hat{v}_m^{(k)}$ coincides with $v_m^{(k)}$ and it satisfies the equation (3.14), which implies that

$$\| \hat{v}_m^{(k)\prime\prime} + L^2(\hat{\varepsilon}_m^{(k)} f^2 - \nu^2)\hat{v}_m^{(k)} \|_{L^2(D_1)} \leq \frac{\text{const}}{L^{2k}} \| \hat{v}_m^{(k)} \|_{L^2(D_1)} \leq \frac{\text{const}}{L^{2k}} \| \hat{v}_m^{(k)} \|_{L^2([0,h])} .$$

Similarly, in D_2 $\hat{v}_m^{(k)}$ coincides with $w_m^{(k)}$ and it satisfies the estimate

$$\| \hat{v}_m^{(k)\prime\prime} + L^2(\hat{\varepsilon}_m^{(k)} f^2 - \nu^2)\hat{v}_m^{(k)} \|_{L^2(D_2)} \leq \frac{\text{const}}{L^{2k}} \| \hat{v}_m^{(k)} \|_{L^2(D_2)} \leq \frac{\text{const}}{L^{2k}} \| \hat{v}_m^{(k)} \|_{L^2([0,h])} .$$

At last, in D_3 $v_m^{(k)}$ is exponentially small because of $\text{Ai}(x)$ is exponentially decreasing as $x \to \infty$, so

$$\| \hat{v}_m^{(k)\prime\prime} + L^2(\hat{\varepsilon}_m^{(k)} f^2 - \nu^2)\hat{v}_m^{(k)} \|_{L^2(D_3)} \leq \frac{\text{const}}{L^{2k}} \| \hat{v}_m^{(k)} \|_{L^2([0,h])} .$$

Since $\hat{v}_m^{(k)}$ is smooth at $s \neq \hat{s}_m^{(k)}$ and $\hat{v}_m^{(k)}$, $\frac{d\hat{v}_m^{(k)}}{ds}$ have no jump at $s = \hat{s}_m^{(k)}$, we get from the last three estimates that

$$\| \hat{v}_m^{(k)\prime\prime} + L^2(\hat{\varepsilon}_m^{(k)} f^2 - \nu^2)\hat{v}_m^{(k)} \|_{L^2([0,h])} \leq \frac{\text{const}}{L^{2k}} \| \hat{v}_m^{(k)} \|_{L^2([0,h])}$$

$$\leq \frac{\text{const}}{L^{2k}} \| f\hat{v}_m^{(k)} \|_{L^2([0,h])}, \tag{4.30}$$

or

$$\| (T - \hat{E}_m^{(k)})\hat{v}_m^{(k)} \| \leq \frac{\text{const}}{L^{2k}} \| \hat{v}_m^{(k)} \| . \tag{4.31}$$

By the general spectral inequality (4.7) it implies that

$$|\hat{E}_m^{(k)} - E_j| \leq \frac{\text{const}}{L^{2k}} \tag{4.32}$$

for some E_j. By (4.28)

$$|E_m^{(k)} - E_j| < \frac{2}{L^{2k-\sigma}}. \tag{4.33}$$

'Differential' part of the proof. Assume that for some quasi-eigenvalue of the zeroth order $E_m^{(0)}$ in the interval (3.29) (recall that we suppose $\rho = 0$) we have an eigenvalue E of the problem (1.1), (1.2) such that

$$|E_m^{(0)} - E| \leq L^{0.1}. \tag{4.34}$$

Let v be the eigenfunction with the eigenvalue E. We shall prove that the number of zeroes of v is not less than $m - 2$ and is not greater than $m + 3$.

Consider the quasi-eigenfunctions $v_- \equiv v_{m-1}^{(0)}$ and $v_+ \equiv v_{m+1}^{(0)}$. The functions v, v_-, v_+ satisfy the equations

$$v'' + qv = 0,$$

$$v_\pm'' + q_\pm v_\pm = 0,$$

where

$$q = Ef^2 - n^2,$$

$$q_\pm = E_{m\pm1}^{(0)} f^2 - n^2 + R_{m\pm1,0}.$$

Sinse $|E_{m\pm1}^{(0)} - E_m^{(0)}| \sim \text{const } L$ and $|R_{m\pm1,0}| \leq \text{const}$, (4.34) implies that

$$q_- < q < q_+.$$

Moreover, let $q_\pm(a_\pm) = q_\pm(b_\pm) = q(a) = q(b) = 0$. Then

$$[a_-, b_-] \subset [a, b] \subset [a_+, b_+].$$

We shall use the following lemma.

Lemma 4.1. *Between any two successive zeroes* $s_l, s_{l+1} \in [a, b]$ *of* v *at least one zero* s_k^+ *of* v_+ *lies. Similarly, between any two successive zeroes* $s_l^-, s_{l+1}^- \in [a_-, b_-]$ *of* v_- *at least one zero* s_k *of* v *lies. Besides,* v *has at most one zero in* $[b, a] \equiv \{s \in S^1 \mid s \leq a \quad or \quad s \geq b\}$.

Proof: Put

$$\frac{v}{v'} = \tan \phi, \quad \phi(s_l) = 0, \quad \phi(s_{l+1}) = \pi,$$

$$\frac{v_+}{v_+'} = \tan \phi_+, \quad 0 \leq \phi_+(s_l) < \pi.$$

Then ϕ, ϕ_+ satisfy Eqs. (4.15), so

$$\phi(s) < \phi_+(s) \quad \text{if} \quad s_l < s,$$

so $\pi < \phi_+(s_{l+1})$, so $\phi_+(s') = 0$ for some $s' \in (s_l, s_{l+1})$, so $v_+(s') = 0$, which was stated. The second statement of Lemma is proved in the same way. Prove the third one. Assume that

$$v(s') = v(s'') = 0, \quad s', s'' \in (b, a) \equiv S^1 \setminus [a, b].$$

Since $q < 0$ in the segment $[s', s'']$,

$$0 = \int_{s'}^{s''} (v'' + qv)v\,dv = \int_{s'}^{s''} (-v'^2 + qv^2)dv < 0.$$

This contradiction completes the proof of Lemma 4.1.

By the construction $v_{m-1}^{(0)}$ has $(m-1)$ zeroes in $[a_-, b_-] \subset [a, b]$, therefore Lemma 4.1 implies that v has at least $(m-2)$ zeroes. Similarly, $v_{m+1}^{(0)}$ has by the construction $(m+1)$ zeroes in $[a_+, b_+] \supset [a, b]$, so by Lemma 4.1 v has at most $(m+2)$ zeroes in $[a, b]$. Since v has at most one zero in (b, a) the total number of zeroes of v does not exceed $(m+3)$, which was stated. This finishes the differential part of the proof.

Thus we have proved that

(i) In $L^{0.1}$-neighborhood of each quasi-eigenvalue $E_m^{(0)}$ in the interval (3.29) at least one eigenvalue E_j lies;

(ii) The number of zeroes of any eigenfunction v_j such that $|E_j - E_m^{(0)}| \leq L^{0.1}$ is between $m-2$ and $m+3$.

Let $E_{m_0}^{(0)}$ be the minimal quasi-eigenvalue in the interval (3.29) and $E_{m_1}^{(0)}$ be the maximal one. Then by (4.20)

$$m_0 = [L\mu_*(\lambda) - \frac{1}{2}] + 1,$$

$$m_1 = [L\mu^*(\lambda) - \frac{1}{2}],$$

so $m_1 - m_0 \sim \text{const} \, L$. Consider the eigenvalues E_{j_0}, E_{j_1} such that

$$|E_{m_0}^{(0)} - E_{j_0}| < L^{0.1},$$

$$|E_{m_1}^{(0)} - E_{j_1}| < L^{0.1}.$$

Then v_{j_0} has at least $m_0 - 2$ zeroes, so by Theorem 2

$$j_0 \geq m_0 - 3.$$

Similarly, v_{j_1} has at most $m_1 + 3$ zeroes, so

$$j_1 \leq m_1 + 3.$$

It implies that near each $E_m^{(0)}$ in the interval (3.29) except at most 6 of them exactly one eigenvalue E_j, $j = j(m)$, lies. Let us show that $j(m) = m$.

Consider arbitrary $E_m^{(0)}$ near which exactly one E_j lies. We have:

$$(T - E_j)v_j = 0,$$

and by (4.31), (4.32)

$$\| (T - E_j)v \| \leq \frac{\text{const}}{L^{2k}},$$

where

$$v = \frac{\hat{v}_m^{(k)}}{\| \hat{v}_m^{(k)} \|}.$$

Let

$$v = \alpha v_j + v_\perp,$$

where $(v_j, v_\perp) = 0$. Then

$$\| (T - E_j)v_\perp \| = \| (T - E_j)v \| \leq \frac{\text{const}}{L^{2k}}.$$

On the other hand by the general spectral inequality

$$\| (T - E_j)v_\perp \| \geq \text{dist}(E_j, \{E_i, i \neq j\}) \cdot \| v_\perp \| \geq \text{const } L \| v_\perp \|,$$

so we get that

$$\| v_\perp \| \leq \frac{\text{const}}{L^{2k+1}}.$$

It implies that

$$\| v_\perp \|_{L^2(S^1)} \leq \frac{\text{const}}{L^{2k+1}}. \tag{4.35}$$

Moreover, since

$$\| v_\perp'' + (E_j f^2 - n^2)v_\perp \|_{L^2(S^1)} = \| (T - E_j)v_\perp \| \leq \frac{\text{const}}{L^{2k}},$$

we get that

$$\| v_\perp'' \|_{L^2(S^1)} \leq \frac{\text{const}}{L^{2k-1}}. \tag{4.36}$$

The inequalities (4.35), (4.36) implies that

$$\| v_\perp \|_{C^1(S^1)} \leq \frac{\text{const}}{L^{2k-1}}. \tag{4.37}$$

Now, $v = \hat{v}_m^{(k)}$ has in $[a, b]$ m zeroes and it is strongly oscillating there:

$$|v'| = |\frac{1}{\sqrt{\varphi'}}\text{Ai}'(-L^{2/3}\varphi)L^{2/3}(-\varphi') + ...| \geq \text{const } L^{2/3}$$

at the zero points, so (4.37) implies that $\alpha v_j = v - v_\perp$ and hence v_j have also m zeroes in $[a, b]$.

Now, according to Lemma 4.1 v_j has at most one zero outside of $[a, b]$, so the total number of its zeroes is equal to m or $m + 1$, but it is even so it is equal to $2[(m+1)/2]$. By Theorem 2 it implies that $j = m$. Thus we have proved that for all m between $m_*(\lambda)$ and $m^*(\lambda)$ except maybe 6 of them

$$|E_m^{(k)} - E_m| \leq \frac{1}{L^{2k-\sigma}}. \tag{4.37}$$

Remark that $E_m^{(k)}$ and E_m are increasing with m and $E_{m+1}^{(k)} - E_m^{(k)} \sim \text{const } L$. It implies that those 6 (at most) exceptional m's are extreme, i.e. they lie near m_* or m^*. Changing λ by $\lambda/2$ in the definition of m_*, m^* we get (4.37) also for them. Theorem 4 is proved.

5. The Problem of Quantum Chaos

In [1] a beautiful geometrical problem was considered which enables to study the distribution of quasi-eigenvalues of the Laplace operator on 2-dimensional revolution surface. In present Section we specify a little the problem to consider the distribution of eigenvalues of this operator. The main specifications concern the following two aspects of the problem:

(i) To consider the distribution of eigenvalues it is neccessary to take into account not only the zeroth term of the QC expansion but also the first one because it is of order of 1;

(ii) The quantization rules are somewhat different in the absence of the turning points and in their presence and it has some consequences for the eigenvalue distribution.

Recall some definitions from [1]. Let $f(x)$ be a smooth periodic function of $x \in \mathbb{R}^1$, $f(x + h_0) = f(x)$, $h_0 > 0$, and S be the revolution surface which arises when the graph of function $f(x)$ on the segment $[0, h_0]$ is rotated around the x-axis, the points $(0, \varphi)$ being identified with (h_0, φ), $0 \leq \varphi \leq 2\pi$. Topologically S is a torus. Consider a Riemannian metrics on S,

$$ds^2 = f^2(x)d\varphi^2 + (1 + (f'(x))^2)dx^2,$$

which is induced by the Euclidean metrics in $\mathbb{R}^3$. The Laplace-Beltrami operator on S is

$$\Delta u = \frac{1}{f\sqrt{1 + (f')^2}} \frac{\partial}{\partial x} \left(\frac{f}{\sqrt{1 + (f')^2}} \frac{\partial u}{\partial x} \right) + \frac{1}{f^2} \frac{\partial^2 u}{\partial \varphi^2}.$$

To simplify this formula introduce a new variable $s = s(x)$ such that

$$\frac{ds}{dx} = \frac{\sqrt{1 + (f')^2}}{f}, \quad \text{or} \quad s = \int_0^x \frac{\sqrt{1 + (f')^2}}{f} dx'.$$

Then we get

$$\Delta u = \frac{1}{f^2} \left(\frac{\partial^2 u}{\partial s^2} + \frac{\partial^2 u}{\partial \varphi^2} \right),$$

$$u(s + h, \varphi) = u(s, \varphi), \quad h = s(h_0).$$

The equation for the eigenfunctions of the operator $-\Delta$ has the form $-\Delta u = Eu$, or

$$-\left(\frac{\partial^2 u}{\partial s^2} + \frac{\partial^2 u}{\partial \varphi^2} \right) = Ef^2 u.$$

Look for u in the form

$$u(s, \varphi) = v(s) \exp(in\varphi).$$

Then we get for $v(s)$ the equation

$$\frac{d^2 v}{ds^2} + (Ef^2 - n^2)v = 0,$$

which coincides with (1.1). The problem of quantum chaos, or maybe it is better to say of 'quantum order' since we started with the integrable classical system, is connected with the study of the distribution of the eigenvalues $E = E_{mn}$, when $E_{mn} \to \infty$. The QC formulae for E_{mn} enable to reduce this study to a geometrical problem.

Consider for any $\mu, \nu \geq 0$ the equation

$$\frac{1}{\pi} \int_a^b \sqrt{\varepsilon f^2(s) - \nu^2} ds = \mu, \tag{5.1}$$

where a, b are the turning points,

$$\varepsilon f^2(s) - \nu^2|_{s=a,b} = 0, \quad \text{if} \quad \mu \leq k^*\nu, \tag{5.2}$$

and

$$a = 0, \quad b = h, \quad \text{if} \quad \mu \geq k^*\nu, \tag{5.2'}$$

where

$$k^* = \frac{1}{\pi} \int_a^b \sqrt{\frac{f^2(s)}{f_{min}^2} - 1} \, ds. \tag{5.3}$$

As concerns the solvability of Eq. (5.1), note that the function

$$I(\varepsilon) \equiv \frac{1}{\pi} \int_a^b \sqrt{\varepsilon f^2(s) - \nu^2} \, ds$$

is increasing,

$$I'(\varepsilon) > 0 \quad \text{for} \quad \varepsilon > \varepsilon_*,$$

and $I(\varepsilon_*) = 0$, $I(\infty) = \infty$, so Eq. (5.1) has a unique solution $\varepsilon = \varepsilon(\mu, \nu)$. One can see easily from Eq. (5.1) that

$$\varepsilon(t\mu, t\nu) = t^2 \varepsilon(\mu, \nu) \quad \forall t > 0. \tag{5.4}$$

Consider the level lines of the function $\varepsilon(\mu, \nu)$,

$$\gamma_R = \{\mu, \nu | \varepsilon(\mu, \nu) = R^2\}.$$

Eq. (5.4) implies that

$$\gamma_R = R\gamma_1. \tag{5.5}$$

Remark, that γ_1 is given by the equation

$$\frac{1}{\pi} \int_a^b \sqrt{f^2(s) - \nu^2} \, ds = \mu,$$

i.e. it is the graph of the function

$$\mu = \mu(\nu) \equiv \frac{1}{\pi} \int_a^b \sqrt{f^2(s) - \nu^2} \, ds.$$

The function $\mu(\nu)$ possesses the following properties:

(i) $\mu(\nu)$ is even;

(ii) $\mu'(\nu) < 0$ if $\nu > 0$;

(iii) $\mu(0) = \frac{1}{\pi} \int_a^b f(s) \, ds$, $\mu'(0) = 0$, $\mu(f_{\max}) = 0$, $\mu'(f_{\max}) = -\sqrt{\frac{f_{\max}}{f''(s_{\max})}} < 0$;

(iv) $\mu(\nu)$ is smooth at $\nu \neq f_{\min}$ and $\mu''(\nu) \sim C_{\pm}|\nu - f_{\min}|^{-1/2}$ as $\nu \to f_{\min} \pm 0$.

Following [1] let us write γ_1 in the polar coordinates r, α, $\mu = r\cos\alpha$, $\nu = r\sin\alpha$:

$$\gamma_1 = \{r, \alpha \mid r = G(\alpha)\}.$$

Then

$$\gamma_R = R\gamma_1 = \{r, \alpha \mid r = RG(\alpha)\}.$$

Put

$$\alpha^* = \mathrm{artan}\, k^* = \mathrm{artan}\left\{ \frac{1}{\pi} \int_a^b \sqrt{\frac{p^2(s)}{f_{\min}^2} - 1} \, ds \right\}. \tag{5.6}$$

Then according to (5.2), (5.2'), (5.3) the condition of the existence of the turning points is $\alpha > \alpha^*$.

Fix now $E, c > 0$ and consider eigenvalues E_{mn} in the interval $E \leq E_{mn} < E + c$. Denote $\alpha(m, n) = \mathrm{artan}\frac{n}{m}$. The QC formulae for E_{mn} are a little bit different for $\alpha(m, n) < \alpha^*$ and $\alpha(m, n) > \alpha^*$. Introduce the sectors

$$V_\delta^+ = \{r, \alpha \mid \frac{\pi}{2} - \delta > \alpha > \alpha^* + \delta\},$$

$$V_\delta^- = \{r, \alpha \mid \alpha^* - \delta > \alpha > 0\}, \quad \delta > 0,$$

$$V^\pm = V_\delta^\pm|_{\delta=0}.$$

Let

$$\xi_\delta^\pm(E, c) = \sharp\{E_{mn} \mid (m, n) \in V_\delta^\pm, \ E \leq E_{mn} \leq E + c\},$$

$$\xi^\pm = \xi_\delta^\pm|_{\delta=0},$$

where $\sharp X$ denotes the number of elements of X. Fix two positive numbers $a_1, a_2, \ a_1 < a_2$ and consider the sets

$$A_{k,\delta}^{\pm}(L,c) = \{R \mid a_1 L \le R \le a_2 L, \quad \xi_\delta^{\pm}(R^2, c) = k\},$$

$$A_k^{\pm}(L,c) = A_{k,\delta}^{\pm}(L,c)|_{\delta=0}.$$

Put

$$p_{k,\delta}^{\pm}(L,c) = \frac{l(A_{k,\delta}^{\pm}(L,c))}{L(a_2 - a_1)},$$

$$p_k^{\pm}(L,c) = p_{k,\delta}^{\pm}|_{\delta=0},$$

where $l(X)$ denotes the Lebesque measure of $X \subset \mathbb{R}^1$.

In [6] some convincing (although non-rigorous) arguments were given in the favor of the following general conjecture: If the underlying classical system is integrable then the eigenvalues of the quantum system in large spectral intervals $a_1 L < \sqrt{E} < a_2 L$ in the limit $L \to \infty$ are in some sense independent and they can be considered as a realization of a Poisson process. Different exact formulations of this conjecture are possible. In [1] for the model under consideration it was formulated as a statement on the limit Poisson distribution of the random variable $\xi(E, c) = \sharp\{E_{mn} \mid E \le E_{mn} \le E + c\}$, where $\sqrt{E}$ is uniformly distributed on the segment $[a_1 L, a_2 L]$, when $L \to \infty$. It is neccessary to make here several remarks.

Remark first that by (1.1) $E_{mn} = E_{m,-n}$, so each eigenvalue is twice degenerate, so $\xi(E, c)$ takes only even values. Next, since QC expansions are different for $(m, n) \in V^{\pm}$ it is convenient to consider separately the distributions of E_{mn} for $(m, n) \in V^-$ and for $(m, n) \in V^+$, or even for $(m, n) \in V_\delta^-$ and $(m, n) \in V_\delta^+$ taking next the limit $\delta \to 0$. Theorems 3,4 enable to study the distributions of the QC eigenvalues instead of the ones of the eigenvalues themselves.

For $(2m, n) \in V_\delta^-$ with the $O\left(\frac{1}{L}\right)$-precision we have by Theorem 3 (see also (2.34)) that

$$E_{2m-1,n} = E_{2m,n}, \tag{5.7}$$

$$E \le E_{2m,n} \le E + c \Leftrightarrow (2m, n) \in \Sigma_{\sqrt{E}}, \tag{5.8}$$

where

$$\Sigma_R = \left\{ r, \alpha \mid G(\alpha)\left(R + \frac{\zeta(\alpha)}{2R}\right) \le r \le G(\alpha)\left(R + \frac{\zeta(\alpha) + c}{2R}\right)\right\}. \tag{5.9}$$

$$\zeta(\alpha) = g\left(\frac{1}{G(\alpha)}\right). \tag{5.10}$$

Here $r = G(\alpha)$ is the equation of γ_1 and the function $g(\varepsilon)$ was defined in (2.31). Note that (5.7) implies that each eigenvalue in V_δ^- is (asymptotically for $L \to \infty$) four times degenerate, hence $\xi_\delta^-(E, c)$ takes values only of the form $k = 4l$.

In [1] QC eigenvalues of the zeroth order were considered which corresponds to $\zeta(\alpha) \equiv 0$ in (5.9). All the technique of [1,5] can be applied with minor modifications to the study of the distribution of $\#\{(2m, n) \in \Sigma_{\sqrt{E}} \cap V_\delta^-\}$ and it gives that

$$\lim_{k \to \infty} p_{4k,\delta}^-(L, c) = \exp(-\lambda_-)\frac{\lambda_-^k}{k!},$$

$$\lambda_- = \frac{c}{4} \int_0^{\alpha^* - \delta} G^2(\alpha) d\alpha,$$

for any 'typical' curve γ_1 (see [1,5] for exact formulations). Taking the limit $\delta \to 0$ one gets a similar result for the distribution of $\xi^-(E, c)$.

In the sector $(m, n) \in V_\delta^+$ we have by Theorem 4 that with $O\left(\frac{1}{L}\right)$-precision

$$E \le E_{mn} \le E + c \Leftrightarrow (m + \frac{1}{2}, n) \in \Sigma_{\sqrt{E}},$$

where Σ_R is defined by (5.9), (5.10) with $g(\varepsilon)$ in (5.10) given by (3.28). In the sector V_δ^+ each eigenvalue is twice degenerate and the technique of [1,5] enables to prove that

$$\lim_{k \to \infty} p_{2k,\delta}^+(L, c) = \exp(-\lambda_+)\frac{\lambda_+^k}{k!},$$

$$\lambda_+ = \frac{c}{2} \int_{\alpha^* + \delta}^{\frac{\pi}{2} - \delta} G^2(\alpha) d\alpha$$

for any 'typical' curve γ_1. Taking the limit $\delta \to 0$ one obtains similar result for $\delta = 0$.

Thus in the limit $L \to \infty$ we have that $\xi(E, c) \to \xi(c)$ which is the sum of $\xi^-(c)$ and $\xi^+(c)$ where $\xi^-(c)$ takes the values $k = 4l$ and $\Pr\{\xi^-(c) = 4l\} = \exp(-\lambda_-)\frac{\lambda_-^l}{l!}$, $\lambda_- = \frac{c}{4} \int_0^{\alpha^*} G^2(\alpha) d\alpha$, and $\xi^+(c)$ takes values $k = 2l$ and $\Pr\{\xi^+(c) = 2l\} = \exp(-\lambda_+)\frac{\lambda_+^l}{l!}$, $\lambda_+ = \frac{c}{2} \int_{\alpha^*}^{\pi/2} G^2(\alpha) d\alpha$. It is worth to note that various aspects of the spectrum degeneracy in the problem of quantum chaos were discussed in [7].

It is neccessary to mention here that our considerations above had a small defect: We tacitly assumed that there is a distribution of random functions $f(x)$ which secures the fulfilment of the conditions 1°-3° of [1] on the random curve γ_1. We think that it is true but at the moment we have not full proof of this assumption.

Remark finally that in [8] some properties of the eigenvalues distribution were studied for revolution surfaces which are topologically isomorphic to sphere. An interesting problem is to extend the results of [1,5] (with some specifications of the present paper) to that case. Note, in particular, that for that case the segment $[0, h]$ is substituted by the whole axis $\mathbb{R}^1$ and the spectral problem (1.1) is stated on infinite line. Moreover turning points always exist in that case and so the quantization conditions have the form (3.25), (3.27).

Acknowledgements. The author thanks Ya. G. Sinai for useful discussions of the problem under consideration and for giving the manuscripts of Refs. [1,5] prior publication. He is also grateful to Italian C.N.R. -G.N.F.M. for the financial support given to his visit to the University of Rome "La Sapienza", where the main part of this work was done. He thanks his Italian friends A. Pellegrinotti, E. Presutti, C. Boldrigini and others for kind hospitality during the visit.

References

1. Ya.G. Sinai, Mathematical problems in the theory of quantum chaos, (1990), to appear.
2. R. Courant and D. Hilbert, Methods of Mathematical Physics, Vol. I, Intersci. Publs. Inc., J. Wiley & Sons, New York e.a. (1970).
3. F.V. Atkinson, Discrete and Continuous Boundary Problems, Academic Press, New York-London (1964).
4. L.D. Landau and Lifschitz, Quantum Mechanics. Non-Relativistic Theory, Pergamon Press, London-Paris (1958).
5. Ya.G. Sinai, Poisson distribution in a geometrical problem, Advances in Soviet Mathematics, Publications of AMS, to appear.
6. M.V. Berry and M. Tabor, Level clustering in the regular spectrum, Proc. Roy. Soc. A 356 (1977), 375-394.
7. J. Keating and R. Mondragon, Quantum chaology of energy levels. Notes based on lectures by Michael Berry, in "Nonlinear Evolution and Chaotic Phenomena", NATO ASI Series B: Physics 176, eds. G. Gallavotti and P.W. Zweifel, Plenum Press, New York-London (1988), 189-196; and M.V. Berry, Aspects of degeneracy, in "Chaotic Behavior in Quantum Systems", NATO ASI Series B: Physics 120, Eds. G. Casati, Plenum Press, New York-London (1985), 123-140.
8. R. Balian and C. Bloch, Distribution of Eigenfrequencies for the wave equation in a Finite Domain: III. Eigenfrequency Density Oscillations, Annals of Physics 69 (1972), 76-160.

A STRENGTHENED ISOPERIMETRIC INEQUALITY FOR SIMPLICES

Alexander G. Reznikov*

Raymond and Beverly Sackler
Faculty of Exact Sciences
Tel-Aviv University
Tel-Aviv, Israel

The celebrated classical isoperimetric inequality states [1] - [3] that for any convex body B in $\mathbf{R}^n$, $Vol_n(B) \leq C_0(n)(Vol_{n-1}(\partial B))^{\frac{n}{n-1}}$, where $C_0(n) = n^{\frac{n}{1-n}}(\beta(n))^{\frac{1}{1-n}}$ and equality holds iff B is a ball ($\beta(n)$ is the volume of the unit ball D_n). If we restrict ourselves to a certain subclass of convex bodies, we should expect a similar inequality with a smaller constant $C(n)$. This is shown for simplices [3] and the corresponding value of $C(n)$ is $C_1(n) = v(n)[(n+1)v(n-1)]^{\frac{n}{1-n}}$. Equality holds iff the simplex is regular (here $v(n)$ is the volume of the regular simplex with unit edge.) In other words, denoting by K an arbitrary simplex in $\mathbf{R}^n$, its volume by $Vol_n K$ and the $(n-1)$-volumes of its $(n-1)$-dimensional faces by $S_0 \cdots S_n$, we have

$$Vol_n(K) \leq C_1(n)\,(S_0 + \cdots + S_n)^{\frac{n}{n-1}} \tag{1}$$

In this paper we'll show that a much stronger inequality holds, namely

$$Vol_n(K) \leq C_2(n)\,(\frac{S_0 \cdots S_n}{S_0 + \cdots + S_n})^{\frac{1}{n-1}} \tag{2}$$

By the Cauchy inequality $S_0 \cdots S_n \leq \left(\frac{1}{n+1}(S_0 + \cdots + S_n)\right)^{n+1}$, so we obtain from (2) the inequality (1) and

$$Vol_n(K) \leq C_3(n)\,(S_0 \cdots S_n)^{\frac{n}{n^2-1}} \tag{3}$$

The values of constants are $C_2(n) = v(n)\left(\frac{1}{n+1}(v(n-1)^n)\right)^{\frac{1}{1-n}}$ and $C_3(n) = v(n)(v(n-1))^{\frac{n}{1-n}}$. Denoting by $S_j^{(k)}$ volumes of k-dimensional faces at K, we obtain from (3) by induction

$$Vol_n(K) \leq C^{(k)}(n)(\prod_j S_j^{(k)})^{\frac{n}{k \cdot \binom{n+1}{k+1}}} \tag{4}$$

where $C^{(k)}(n) = v(n)\,(v(k))^{-\frac{n}{k \cdot \binom{n+1}{k+1}}}$. This inequality should be looked at as a specialization of the (higher) Alexandrov inequalities [1] for simplices.

---

* Supported in part by the Germany-Israel Foundation (GIF)

The proof of our main inequality (2) is based on some special symmetrization-type construction. Let A_i, $0 \le i \le n$, be the vertices of K and let O be the center of the inscribed sphere. Let B_i, $0 \le i \le n$, be the touching points of this sphere with the face opposite to A_i. Let r be the radius of the inscribed sphere. We're going to compute the volume of the simplex $B_0 \cdots B_n$ and to apply then the well-known inequality [3]

$$Vol_n(B_0 \cdots B_n) \le E(n)r^n , \tag{5}$$

where $E(n) = (\frac{n^2}{n+1})^{-n}(v(n))^{1-n}(v(n-1))^n$.

To compute the left side of (5), we need some preparational work.

Definition 1. Let X be a finite dimensional real vector space with the volume form $\Omega \in \wedge^n X^*$, $n = \dim X$. Let $a_1, \ldots, a_{n-1} \in X$. By the mixed product $[a_1, \ldots, a_{n-1}]$ we denote the linear functional $a \mapsto \Omega[a_1, \ldots, a_{n-1}, a]$. If X is euclidean, we consider $[a_1, \ldots, a_{n-1}]$ to be an element of X.

Lemma 1. Let X be the euclidean n-dimensional space with the canonical volume form Ω and let $a_1, \ldots, a_n \in X$. Then

$$\Omega([a_2, \ldots, a_n], [a_1, a_3, \ldots, a_n], \ldots, [a_1, \ldots, a_{n-1}]) = (\Omega(a_1, \ldots, a_n))^{n(n-1)} \tag{6}$$

Proof: Let A be any linear unimodular operator in X. Then it follows immediately from Definition 1, that for any $b_1, \ldots, b_{n-1} \in X$, $[Ab_1, \ldots, Ab_{n-1}] = (A^*)^{-1}[b_1, \ldots, b_{n-1}]$. Indeed, for $b \in X$ we have $(A^*[Ab_1, \ldots, Ab_{n-1}], b) = ([Ab_1, \ldots, Ab_{n-1}], Ab) = \Omega(Ab_1, \ldots, Ab_{n-1}, Ab) = (\det A)^n (b_1, \ldots, b_{n-1}b) = ([b_1, \ldots, b_{n-1}], b)$, because $\det A = 1$. Now by Gram-Schmidt we can find such A that all Aa_i will be orthogonal to each other, so it is sufficient to verify (6) for orthogonal a_i, which is very easy.

Lemma 2. With the notation of Lemma 1, we have:
$\|[a_1, \ldots, a_{n-1}]\| = Vol_{n-1}K(a_1, \ldots, a_{n-1})$, where $K(a_1, \ldots, a_{n-1})$ is the parallelotope of dimension $(n-1)$, spanned by $a_1, \ldots, a_{n-1}$.

Proof: Choose a to be orthogonal to $a_1, \ldots, a_{n-1}$. Then by definition 1 we have $([a_1, \ldots, a_{n-1}], a_i) = 0$, when $1 \le i \le n-1$, so $[a_1, \ldots, a_{n-1}] = \lambda a$ for some $\lambda \in \mathbb{R}$. As $|([a_1, \ldots, a_{n-1}], a)| = |\Omega(a_1, \ldots, a_{n-1}, a)| = \|a\| Vol_{n-1}K(a_1, \ldots, a_{n-1})$, we have $|\lambda| \|a\|^2 = \|a\| Vol_{n-1}K(a_1, \ldots, a_{n-1})$ and $\|[a_1, \ldots, a_{n-1}]\| = |\lambda| \cdot \|a\| = Vol_{n-1}K(a_1, \ldots, a_{n-1})$.

We return to the computation of the volume of the simplex $B_0 \cdots B_n$. The simplex can be divided naturally into the parts $OB_0 \cdots \hat{B}_j \cdots B_n$. Consider one part, say $OB_1, \ldots, B_n$.

Denote $a_i = \overrightarrow{A_o A_i}$. For $1 \le i \le n$ the vector $\overrightarrow{OB_i}$ is orthogonal to $a_1, \ldots, \widehat{a_i}, \ldots, a_n$, so $\overrightarrow{OB_i} = \lambda_i [a_1, \ldots, \widehat{a_i}, \ldots, a_n]$, for some λ_i. But $\| \overrightarrow{OB_i} \| = r$, so, applying Lemma 2, we see

$$\lambda_i = \frac{r}{Vol_{n-1} K(a_1, \ldots, \widehat{a_i}, \ldots, a_n)}$$

Since, $S_i = \frac{1}{(n-1)!} Vol_{n-1} K(a_1, \ldots, \widehat{a_i}, \ldots, a_n)$, we obtain that $\overrightarrow{OB_i} = (n-1)! [a_1, \ldots, \widehat{a_i}, \ldots, a_n]$. It follows that

$$Vol(OB_1, \ldots, B_n) = \frac{1}{n!} \frac{r^n}{((n-1)!)^n S_1 \cdots S_n} |\Omega([a_2, \ldots, a_n], \ldots, [a_1, \ldots, a_{n-1}])| .$$

By Lemma 1 we can represent the right side by $\frac{r^n}{n!((n-1)!)^n S_1 \cdots S_n} \times |\Omega(a_1, \ldots, a_n)|^{n-1} = \frac{(n!)^{n-1}}{n!((n-1)!)^n} \frac{r^n}{S_1 \cdots S_n} (Vol_n(K))^{n-1}$. Summing over all the parts, we obtain the following

Lemma 3. The volume of the simplex $B_0, \ldots, B_n$ can be expressed as

$$(7) \qquad Vol_n(B_0 \cdots B_n) = \frac{(n!)^{n-2}}{((n-1)!)^n} \frac{r^n (S_0 + \cdots + S_n)}{S_0 \cdots S_n} (Vol_n(K))^{n-1}$$

Proof of the main inequality. Combining (5) and (7) we obtain

$$\frac{S_0 + \cdots + S_n}{S_0 \cdots S_n} (Vol_n(K))^{n-1} \le (\frac{n^2}{n+1})^{-n} (v(u))^{1-n} (v(n-1))^n \frac{(n!)^2}{n^n} ,$$

or

$$Vol_n(K) \le C_4(n) (\frac{S_0 \cdots S_n}{S_0 + \cdots + S_n})^{\frac{1}{n-1}} \qquad (8)$$

where $C_4(n) = v(n)^{-1} (v(n-1))^{\frac{n}{n-1}} \frac{(n+1)^{\frac{n}{n-1}} (n!)^{\frac{2}{n-1}}}{n^{\frac{3n}{n-1}}}$. If the initial simplex $A_0 \cdots A_n$ is regular, then (2) becomes equality. In this case the simplex $B_0 \cdots B_n$ is also regular, hence [2](5) becomes equality, too. It follows that $C_2(n) = C_4(n)$, which proves (2). Let us check what gives us the identity $C_2(n) = C_4(n)$. We have

$$v(n) (\frac{1}{n+1} (v(n-1))^n)^{\frac{1}{1-n}} = v(n)^{-1} (v(n-1))^{\frac{n}{n-1}} \frac{(n+1)^{\frac{n}{n-1}} (n!)^{\frac{2}{n-1}}}{n^{\frac{3n}{n-1}}} ,$$

or

$$\frac{(v(n))^{\frac{2}{n}}}{(v(n-1))^{\frac{2}{n-1}}} = \frac{(n!)^{\frac{2}{n(n-1)}} (n+1)^{\frac{1}{n}}}{n^{\frac{3}{n-1}}} .$$

One can easily derive from this the formula

$$v(n) = \frac{\sqrt{n+1}}{n! 2^{\frac{n}{2}}}$$

expressing explicitly the volume of the regular simplex.

Let us finish this note with two natural questions.

1. (A. Dvoretzky). Is it true, that, by repeating the construction of the inscribed simplex $B_0 \cdots B_n$ from the initial simplex $A_0 \cdots A_n$, we'll obtain a sequence of simplices which tends to a regular one (up to the scale)? For $n = 2$ this is shown in [5].

2. Are there some analogues of (1) and (2) in the hyperbolic n-space? What are the connections between these analogues and known results [4]?

References

[1] Yu. Burago, V. Zalgaller, Geometric inequalities, Springer Verlag, 1988.

[2] L. Fejes Toth, Lagerungen in der Ebene auf der Kugel und in Raum, Springer Verlag, 1957.

[3] H. Hadwiger, Vorlesungen uber Inhalt, Oberflache und Isoperimetrie, Springer Verlag, 1957.

[4] W. Thurston, The geometry and topology of 3-manifolds, Princeton, 1978.

[5] A. Reznikov, Two sequences of triangles, Quant, N8 (1975) (Russian).

A NEW ISOPERIMETRIC INEQUALITY AND
THE CONCENTRATION OF MEASURE PHENOMENON

Michel Talagrand*

U.A. au C.N.R.S. No. 754	and	Department of Mathematics
Equipe d' Analyse-Tour 46		The Ohio State University
Université Paris VI		231 West 18th Avenue
4 Place Jussieu		Columbus, Ohio 43210
75230 Paris Cedex 05		USA
France		

Abstract

We prove a new isoperimetric inequality for a certain product measure that improves upon some aspects of the "large deviation" consequences of the isoperimetric inequality for the Gaussian measure.

1. Introduction

We denote by γ the canonical Gaussian measure on $\mathbb{R}$, of density $\frac{1}{\sqrt{2\pi}} e^{-x^2/2}$ with respect to Lebesgue measure. We denote by γ^∞ the product measure on $\mathbb{R}^{\mathbb{N}}$, where each factor is endowed with γ. Throughout the paper, we set, for $\alpha = 1, 2$,

$$B_\alpha = \{x \in \mathbb{R}^{\mathbb{N}} ; \quad \sum_{k \geq 1} |x_k|^\alpha \leq 1\}.$$

For two sets A, B of $\mathbb{R}^{\mathbb{N}}$, we set

$$A + B = \{x + y; \quad x \in A, \quad y \in B\}.$$

Consider a (Borel) set $A \subset \mathbb{R}^{\mathbb{N}}$, and $a \in \mathbb{R}$ such that $\gamma^\infty(A) = \gamma((-\infty, a])$. The isoperimetric inequality for the Gauss measure states that for $u \geq 0$,

$$\gamma^\infty_*(A + uB_2) \geq \gamma((-\infty, a + u]) \tag{1.1}$$

---

* Work partially supported by an N.S.F. grant.

(the inner measure is needed as $A + uB_2$ might fail to be measurable). This inequality plays a fundamental role in the theory of Gaussian processes and Gaussian measures. It was discovered independently by C. Borell [B] and B.S. Tsirelson and V.N. Sudakov [ST]. They derived this inequality from Lévy's isoperimetric inequality on the sphere, via Poincaré lemma. Later, A. Ehrhard [E1] developed a more intrinsic approach of Gaussian symmetrization that also led him to other remarkable inequalities [E2].

Very often, the Gaussian isoperimetric inequality is used under the following form

$$\gamma^\infty(A) \geq \tfrac{1}{2} \Rightarrow \gamma_*^\infty(A + uB_2) \geq \gamma((-\infty, u]) \qquad (1.2)$$

where the last term is evaluated through the classical estimate $\gamma_1((-\infty, u]) \geq 1 - \tfrac{1}{2}e^{-u^2/2}$. For some applications (e.g. [T]) it is important to have the sharp estimate (1.2). For many others it is sufficient to know, that for some universal constant K, one has

$$\gamma^\infty(A) \geq \frac{1}{2} \Rightarrow \gamma_*^\infty(A + uB_2) \geq 1 - 2e^{-u^2/K}. \qquad (1.3)$$

(The role of the factor 2 being to emphasize that one cares only for large values of u.)

In the terminology of V. Milman [M], (1.3) will be called the concentration of measure property (for the Gaussian measure). The main contribution of the present paper is the somewhat unexpected fact that the concentration of measure property for the Gaussian measure is the consequence of a sharper principle, itself unrelated to Gaussian measure, and that we present now. Throughout the paper, we denote by μ the probability measure on R of density $\tfrac{1}{2}e^{-|x|}$ with respect to Lebesgue measure, and by μ^∞ its product measure on R^N.

Theorem 1.2. *There exists a universal constant K with the following property. Consider a set $A \subset \mathsf{R}^N$, and $a \in \mathsf{R}$ such that $\mu^\infty(A) = \mu((-\infty, a])$. Then*

$$\mu_*^\infty(A + \sqrt{u}B_2 + uB_1) \geq \mu((-\infty, a + \frac{u}{K}]). \qquad (1.4)$$

In particular

$$\mu^\infty(A) \geq \frac{1}{2} \Rightarrow \mu_*^\infty(A + \sqrt{u}B_2 + uB_1) \geq 1 - \frac{1}{2}\exp\left(-\frac{u}{K}\right). \qquad (1.5)$$

A main difference between (1.4) and (1.1) is the fact that A is now enlarged by a mixture of the ℓ^1 and ℓ^2 balls, in proportions that vary with u. Consider a sequence $(t_k)_{k \geq 1}$ (having for simplicity only finitely many non-zero terms), and set $f(x) = \sum_{k \geq 1} t_k x_k$. It is possible to study the tails $\mu^\infty(\{f \geq u\})$ in an elementary way; but we will do it here using (1.5), in order to illustrate the role of the ℓ^1 and ℓ^2 balls.

Consider

$$A = \{x \in \mathbb{R}^\mathbb{N}; \ \sum_{k \geq 1} t_k x_k \leq 0\}.$$

By symmetry, $\mu^\infty(A) \geq \frac{1}{2}$. If $y \in A + \sqrt{u}B_2 + uB_1$, we then have $f(y) \leq \sqrt{u}\|t\|_2 + u\|t\|_\infty$. Thus, by (1.5)

$$\mu^\infty(\{f \geq \sqrt{u}\|t\|_2 + u\|t\|_\infty\}) \leq \frac{1}{2}e^{-u/K}.$$

This implies that

$$\mu^\infty(\{f \geq u\}) \leq \frac{1}{2}e^{-u/K\|t\|_\infty}$$

for $u \geq \|t\|_2^2/\|t\|_\infty$, while, for $0 \leq u \leq \|t\|_2^2/\|t\|_\infty$ we have

$$\mu^\infty(\{f \geq u\}) \leq \frac{1}{2}e^{-u^2/K\|t\|_2^2}.$$

(For the simplicity of notations we allow the value of the constant K to vary at each occurrence.) The exponents in these bounds are exactly of the right order.

Another feature that differentiates (1.4) from (1.1) (and makes it closer to (1.3)) is the constant K on the right. It would be nice (but irrelevant for our purposes) to have a more exact form of (1.4). A natural question to ask in that direction is whether there exists, given u, a natural "smallest" set $W_u \subset \mathbb{R}^\mathbb{N}$ such that

$$\mu_*^\infty(A + W_u) \geq \mu((-\infty, a + u]) \tag{1.6}$$

for all $a \in \mathbb{R}$, all sets $A \subset \mathbb{R}^\mathbb{N}$, such that $\mu^\infty(A) = \mu((-\infty, a])$. The difficulty of that question is that the shape of W_u is likely to vary depending on u. A worthy inequality (1.6) should, in particular, allow to recover excellent tail estimates for $\mu^\infty(\{\sum_{k \geq 1} t_k x_k \geq u\})$; but in view of the variety of the estimates known for this quantity, [H] this does not appear to be a simple task.

We now explain why (1.5) improves upon (1.3). (The basic idea of this argument is due to G. Pisier [P].) Consider the increasing map ψ from $\mathbb{R}$ to $\mathbb{R}$ that transforms μ into γ. It is easy to see that

$$\forall x, y \in \mathbb{R}, \quad |\psi(x) - \psi(y)| \leq K \min(|x - y|, \ |x - y|^{1/2}). \tag{1.7}$$

Consider the map Ψ from $\mathbb{R}^\mathbb{N}$ to itself given by $\Psi((x_k)_{k \geq 1}) = (\psi(x_k))_{k \geq 1}$. It transforms μ^∞ into γ^∞. Consider now a set $A \subset \mathbb{R}^\mathbb{N}$, with $\gamma(A) \geq 1/2$. Then $\mu(\Psi^{-1}(A)) = \gamma(A) \geq 1/2$. By (1.4) we have

$$\mu_*^\infty(\Psi^{-1}(A) + \sqrt{u}B_2 + uB_1) \geq 1 - \frac{1}{2}e^{-u/K}$$

so that

$$\gamma^{\infty}(\Psi(\Psi^{-1}(A) + \sqrt{u}B_2 + uB_1)) \geq 1 - \frac{1}{2}e^{-u/K}.$$

Now using (1.6) it is simple to show that

$$A_u = \Psi(\Psi^{-1}(A) + \sqrt{u}B_2 + uB_1) \subset A + K\sqrt{u}B_2;$$

thus we recover (1.3). It can, however, happen that the set A_u is much smaller than $A + K\sqrt{u}B_2$. A striking example is when A is the cube

$$A = \{x; \quad \forall k \leq n, \quad |x_k| \leq c_n\}$$

where c_n is (say) chosen such that $\gamma^{\infty}(A) = 1/2$, so that c_n is of order $(\log n)^{1/2}$. In that case, and when $u \ll \log n$, it is simple to see that

$$A_u \subset A + K\left(\frac{\sqrt{u}}{\sqrt{\log n}}B_2 + \frac{u}{\sqrt{\log n}}B_1\right) \subset A + K\left(\frac{u}{\sqrt{\log n}}\right)^{1/2}\sqrt{u}B_2.$$

An intriguing aspect of the improvement of (1.3) via (1.4) is that we break the rotational invariance that is fundamental to the Gaussian measure γ^{∞}. Thus (working in $\mathbf{R}^n$ rather than $\mathbf{R}^{\infty}$) if R denotes any rotation of $\mathbf{R}^n$, (with obvious notations) we have $\gamma^n((RA)_u) \geq 1 - \exp(-u/K)$.

Having discovered that (1.3) is a consequence of (1.4), one must now ask whether (1.4) itself is the end of the story. We will show that in the setting of product measures, this seems to be the case. Indeed (Proposition 5.1) if for a measure θ on $\mathbf{R}$, the product measure θ^{∞} on $\mathbf{R}^{\mathbf{N}}$ satisfies even a much weaker form of concentration of measure than (1.4), the function $\theta(\{|x| \geq a\})$ must decrease exponentially fast.

The main difficulty in proving Theorem 1.1 is that, in contrast with the Gaussian case, the measure μ does not have many symmetries. This limits the use of rearrangements. In section 2, we show how an induction argument reduces the proof of Theorem 1.1 to that of a certain statement in dimension 2. A special case of that statement is proved in section 3. In section 4, we use variational arguments to prove that this special case is actually the general case, thereby finishing the proof of Theorem 1.1. While the main ideas of the proof of Theorem 1.1 are rather natural, the proof requires checking a number of tedious facts.

On the other hand, the important part of Theorem 1.1 is certainly (1.5). Fortunately this is much easier to prove. In section 5, we give a simple proof of the following (that is weaker than Theorem 1.1).

Theorem 1.2. *There exists a universal constant K with the following property. Consider a set $A \subset R^N$, and for $x \in R^N$, set*

$$\theta_A(x) = \inf \left\{ u \geq 0 \; ; \; x \in A + \sqrt{u} B_2 + u B_1 \right\} .$$

Then

$$\int^* \exp\left(\theta_A(x)/K\right) d\mu^\infty(x) \leq 2/\mu^\infty_*(A) .$$

We observe that by Markov's inequality, this gives

$$\mu^\infty(A) \geq \tfrac{1}{2} \Rightarrow \mu^\infty_*\left(\{\theta_A(x) \geq Ku\}\right) \leq 4\exp(-u) ,$$

so that

$$\mu^\infty_*\left(A + \sqrt{u} B_2 + u B_1\right) \geq 1 - 4\exp\left(-\frac{u}{K}\right) .$$

For u large enough, this is equivalent to (1.5).

The reader is certainly advised to read the proof of Theorem 1.2 first, going back to previous sections whenever necessary.

2. Induction

We denote by $\varphi(x) = \tfrac{1}{2} e^{-|x|}$ the density of μ with respect to Lebesgue measure. We set

$$\Phi(x) = \int_{-\infty}^{x} \varphi(u) \, du = \mu([-\infty, x]) .$$

We set $\Phi(-\infty) = 0, \Phi(+\infty) = 1$.

The sets $\sqrt{u} B_2 + u B_1$ of Theorem 1.1 are not easy to manipulate. Our first task will be to replace them by more amenable sets.

Consider a parameter $L > 0$, to be determined later. For $x \in R$, we set $\xi(x) = \frac{1}{L} \int_0^{|x|} \frac{u}{1+u} \, du$. There is no magic in this formula. The properties of ξ we really need are that $L\xi(x)$ resembles x^2 for $|x| \leq 1$ and $|x|$ for $|x| \geq 1$, and moreover (for technical reasons) that $\xi'(x)$ is strictly increasing for $x > 0$.

We now define, for $u \geq 0$,

$$V(u) = \{y \in R^N ; \; \sum_{k \geq 1} \xi(y_k) \leq u\}; \; V_n(u) = \{y \in R^n ; \; \sum_{1 \leq k \leq n} \xi(y_k) \leq u\} .$$

We show that

$$V(u) \subset \sqrt{4Lu} \, B_2 + 4Lu B_1 . \tag{2.0}$$

Indeed, consider $y \in V(u)$. Define $z = (z_k) \in \mathsf{R}^N$ by $z_k = y_k$ if $|y_k| \leq 1$, $z_k = 0$ otherwise. Since $\xi(x) \geq x^2/4L$ for $|x| \leq 1$, we have $z \in \sqrt{4Lu}\, B_2$. Since $\xi(x) \geq |x|/4L$ for $|x| \geq 1$, we have $\|y - z\|_1 \leq 4Lu$. This proves (2.0).

To prove Theorem 1.1, it suffices to prove the following.

Theorem 2.1. *One can determine L with the following property. For all $A \subset \mathsf{R}^N$,*

$$\mu^\infty(A) = \Phi(a) \Rightarrow \mu_*^\infty(A + V(u)) \geq \Phi(a + u).$$

(One can then take the constant K of Theorem 1.1 equal to $4L$.) Denote now by μ^n the power of μ on R^n. By an obvious approximation argument, it suffices to prove that we can find L such that the following holds:

(I_n) for each compact set A for R^n, and $u \geq 0$,

$$\mu^n(A) = \Phi(a) \Rightarrow \mu^n(A + V_n(u)) \geq \Phi(a + u).$$

This statement will be proved by induction on n. The first task is to prove the case $n = 1$. In that case, one has actually a much more accurate result (that is the exact analogue of (1.1)).

Proposition 2.2. *Consider a compact set $A \subset \mathsf{R}$. Then*

$$\mu(A) = \Phi(a) \Rightarrow \mu(A + [-u, u]) \geq \Phi(a + u).$$

The proof uses rearrangements. We first consider the case where A is an interval $[v, w]$.

Lemma 2.3. *Consider $v \leq w$, $v + w \leq 0$. Consider $v' \leq v$ (possibly $v' = -\infty$) and w' such that $\mu([v, w]) = \mu([v', w'])$. Then $\mu([v - u, w + u]) \geq \mu([v' - u, w' + u])$.*

Proof: Define the function $y = y(x)$ by

$$\Phi(y) - \Phi(x) = \mu([x, y]) = \mu([v, w]).$$

Thus $y'\varphi(y) = \varphi(x)$. Set now

$$h(x) = \mu([x - u, y + u]) = \Phi(y + u) - \Phi(x - u).$$

Thus

$$h'(x) = y'\varphi(y + u) - \varphi(x - u)$$
$$= \frac{1}{\varphi(y)}[\varphi(x)\varphi(y + u) - \varphi(y)\varphi(x - u)]$$

has the same sign as

$$e^{-|x|-|y+u|} - e^{-|y|-|x-u|}.$$

But we have, for $u > 0$

$$\text{the function} \quad x \rightarrow |x + u| - |x| \quad \text{increases}. \tag{2.1}$$

Since y is obviously an increasing function of x, we have $x + y \leq v + w \leq 0$ for $x \leq v$, so that $y \leq -x$. Thus, by (2.1),

$$|y + u| - |y| \leq |-x + u| - |x|$$

i.e.,

$$|y + u| + |x| \leq |y| + |x - u|.$$

Thus $h'(x) \leq 0$ for $x \leq v$. This proves the result (observe that when $v' = -\infty$, we have $\lim_{x \to -\infty} y(x) = w'$). $\qquad\square$

Lemma 2.4. *Consider* $v \leq w$, *and* z *such that* $\Phi(v) + \Phi(w) = \Phi(z)$. *Then* $\Phi(v+u) + \Phi(w+u) \geq \Phi(z+u)$.

Proof: This could be deducted from the previous lemma, although a direct argument is simpler. We define $y = y(x)$ by $\Phi(y) + \Phi(x) = \Phi(z)$, so that $y'\varphi(y) = -\varphi(x)$. We set $h(x) = \Phi(y + u) + \Phi(x + u)$, so that $h'(x) = y'\varphi(y + u) + \varphi(x + u)$ has the sign of

$$-\varphi(x)\varphi(y + u) + \varphi(y)\varphi(x + u) = e^{-|x+u|-|y|} - e^{-|x|-|y+u|}.$$

For $x \leq v$, we have $y \geq w$, so that $y \geq w \geq v \geq x$. By (2.1) we have $|x + u| - |x| \leq |y + u| - |y|$, i.e. $|x + u| + |y| \leq |x| + |y + u|$, so that $h' \geq 0$. The result follows by letting $x \to -\infty$, since $y \to z$. $\qquad\square$

We now prove Proposition 2.2. It suffices to consider the case where A is a finite union of disjoint intervals. The proof proceeds by induction over the number of bounded intervals of A. When $n = 0$, either A consists of one unbounded interval, and there is nothing to prove, or else it consists of two such intervals, and the result follows from Lemma 2.4.

For the induction step from n to $n + 1$, we can write $A = B \cup I$, where $I = [v, w]$ and where B is a union of intervals, at most n of which are bounded. If

$$(B + [-u, u]) \cap (I + [-u, u]) \neq \emptyset, \tag{2.2}$$

we could replace I and one of the intervals of B by one single interval containing both, without increasing $A + [-u, u]$, thus the result holds in that case. Suppose now that

$$(B + [-u, u]) \cap (I + [-u, u]) = \emptyset$$

and suppose, for definiteness, that $v \leq w$. We set

$$v' = \inf\{x \geq 0, \quad \exists y, \quad \mu([x, y]) = \mu(I), \; [x - u, y + u] \cap (B + [-u, u]) = \emptyset\}$$

and we denote by w' the value of y corresponding to $x = v'$. If we replace I by $[v', w']$, we do not change $\mu(B)$, but, Lemma 2.3 shows that we decrease $\mu(B + [-u, u])$. If $v' = -\infty$, $[v', w']$ is unbounded and we have reduced to the case of n bounded intervals. If $v' > -\infty$, we have reduced to the case (2.2).   □

We can and do assume $L \geq 1$. Then $\xi(x) \leq |x|$, and thus $V_1(u) \supset [-u, u]$, so that Proposition 2.1 shows that (I_1) holds.

Set $\overline{\mathsf{R}} = \mathsf{R} \cup \{-\infty, \infty\}$. For a (Borel) function $f : \mathsf{R} \to \overline{\mathsf{R}}$, we set

$$\Phi(f) = \int_{\mathsf{R}} \Phi(f(x)) \, d\mu(x) = \mu^2(\{(x, y) \in \mathsf{R}^2 \, ; \; y \leq f(x)\}).$$

In Sections 3 and 4, we will prove the following fact.

Proposition 2.5. *The parameter L can be chosen such that the following holds. Consider a non-decreasing function $f : \mathsf{R} \to \overline{\mathsf{R}}$. Consider $u \geq 0$, and set*

$$\overline{f}(x) = \sup\{f(x) + u - \xi(x - x'); \quad \xi(x - x') \leq u\}. \tag{2.3}$$

Then

$$\Phi(f) = \Phi(a) \Rightarrow \Phi(\overline{f}) \geq \Phi(a + u).$$

The rest of that section is devoted to prove that Proposition 2.5 implies that $(I_n) \Rightarrow (I_{n+1})$, thereby proving Theorem 2.1. The first task is to show that Proposition 2.5 implies a similar result when f is no longer assumed to be non-decreasing. This follows from the next result where, as well as in the rest of the paper, for a function f, we denote by $\overline{f}$ the function given by (2.3) (the value of u being fixed).

Proposition 2.6. *Consider a (Borel) function $g : \mathsf{R} \to \overline{\mathsf{R}}$. Define its non-decreasing rearrangement $f : \mathsf{R} \to \overline{\mathsf{R}}$ by*

$$f(-x) = \sup\{y; \quad \mu(\{g \geq y\}) \leq \Phi(x)\}. \tag{2.4}$$

Then f is non-decreasing, $\Phi(f) = \Phi(g)$, $\Phi(\overline{f}) \leq \Phi(\overline{g})$.

Proof: For notational convenience, we will show instead that the (non-increasing) function given by

$$f(x) = \sup\{y; \quad \mu(\{g \geq y\}) \leq \Phi(x)\}$$

satisfies $\Phi(f) = \Phi(g)$; $\Phi(\overline{f}) \leq \Phi(\overline{g})$.

To prove that $\Phi(f) = \Phi(g)$, we have to prove that the subgraphs of f and g have the same measure for μ^2. By Fubini theorem, it suffices to show that for all y, $\mu(\{f \geq y\}) = \mu(\{g \geq y\})$.

Consider x such that $\Phi(x) = \mu(\{g \geq y\})$. Then $f(x) \geq y$ by definition, so that $\{f \geq y\} \supset (-\infty, x]$, and

$$\mu(\{f \geq y\}) \geq \mu((-\infty, x]) = \Phi(x) = \mu(\{g \geq y\}).$$

Consider now x such that $\Phi(x) > \mu(\{g \geq y\})$. Then $f(x) < y$, so that $\{f \geq y\} \subset [-\infty, x]$, and $\mu(\{f \geq y\}) \leq \Phi(x)$. Thus $\mu(\{f \geq y\}) \leq \mu(\{g \geq y\})$.

To prove that $\Phi(\overline{f}) \leq \Phi(\overline{g})$ it suffices, using Fubini theorem again, to show that for all y we have $\mu(\{\overline{f} \geq y\}) \leq \mu(\{\overline{g} \geq y\})$ or, equivalently, $\mu(\{\overline{f} > y\}) \leq \mu(\{\overline{g} > y\})$. Define b by $\Phi(b) = \mu(\{\overline{f} > y\})$. Since f, and hence $\overline{f}$, is non-increasing, we have $\overline{f}(x) > y$ for $x < b$. Consider now $x' \leq x < b$. Set $s = x - x'$, and assume that $\xi(s) \leq u$. Since f is non-increasing, we have

$$\Phi(x') \leq \mu(\{f \geq f(x')\}) = \mu(\{g \geq f(x')\}).$$

By Proposition 2.2, we have

$$\Phi(x) = \Phi(x' + s) \leq \mu(\{g \geq f(x')\} + [-s, s]).$$

For $g(v) \geq f(x')$ and $|w| \leq s$, we have

$$\overline{g}(v + w) \geq g(v) + u - \xi(w) \geq f(x') + u - \xi(s).$$

Thus

$$\Phi(x) \leq \mu(\{\overline{g} \geq f(x') + u - \xi(x - x')\}). \tag{2.5}$$

Since f is non-increasing, we have

$$\overline{f}(x) = \sup\{f(x') + u - \xi(x - x'); \quad x' \leq x, \quad \xi(x - x') \leq u\}.$$

Taking the supremum in (2.5) over x' gives

$$\Phi(x) \le \mu(\{\overline{g} > \overline{f}(x)\}) \le \mu(\{\overline{g} > y\})$$

and thus, taking the supremum over $x < b$ we have $\mu(\{\overline{f} > y\}) \le \mu(\{\overline{g} > y\}))$. $\square$

We now prove the implication $(I_n) \Rightarrow (I_{n+1})$. Assuming that (I_n) holds, consider a compact subset A of $\mathbb{R}^{n+1}$. For $x \in \mathbb{R}$, we denote by A_x the set $\{z \in \mathbb{R}^n \, ; \quad (z, x) \in A\}$.

Step 1.   For $x, x' \in \mathbb{R}$, such that $\xi(x - x') \le u$, we have

$$A_{x'} + V_n(u - \xi(x - x')) \subset (A + V_{n+1}(u))_x \, .$$

Indeed, consider $v \in V_n(u - \xi(x - x'))$. By definition,

$$\sum_{k \le n} \xi(v_k) \le u - \xi(x - x'),$$

and thus $(v, x - x') \in V_{n+1}(u)$. Consider now $z \in A_{x'}$, so that $(z, x') \in A$. We have

$$(z + v, x) = (z, x') + (v, x - x') \in A + V_{n+1}(u)$$

i.e., $z + v \in (A + V_{n+1}(u))_x$.

Step 2.   Define $g(x) = \Phi^{-1}(\mu^n(A_x))$. By Fubini theorem, we have

$$\Phi(g) = \int_{\mathbb{R}} \Phi(g(x)) \, d\mu(x) = \int_{\mathbb{R}} \mu^n(A_x) \, d\mu(x) = \mu^{n+1}(A) = \Phi(a) \, .$$

By Fubini theorem again, we have

$$\mu^{n+1}(A + V_{n+1}(u)) = \int_{\mathbb{R}} \mu^n((A + V_{n+1}(u))_x) \, d\mu(x) \, .$$

Step 1 shows that, whenever $\xi(x - x') \le u$, we have

$$(A + V_{n+1}(u))_x \supset A_{x'} + V_n(u - \xi(x - x'))$$

so that, by induction hypothesis (I_n), we have

$$\mu((A + V_{n+1}(u))_x) \ge \Phi(g(x') + u - \xi(x - x'))$$

and, taking the supremum over x',

$$\mu^n((A + V_{n+1}(u))_x) \ge \Phi(\overline{g}(x)) \, .$$

Thus

$$\mu^{n+1}(A + V_{n+1}(u)) \ge \int_{\mathbb{R}} \Phi(\overline{g}(x)) d\mu(x) = \Phi(\overline{g}) \, ,$$

and the conclusion follows from Propositions 2.5, 2.6. $\square$

3. Basic inequality

The aim of this section is to prove the following.

Proposition 3.1. *It is possible to choose the parameter L such that the following holds. Consider a non-decreasing function $f : \mathsf{R} \to \overline{\mathsf{R}}$, and set*

$$\widehat{f}(x) = \sup_{y \in \mathsf{R}} f(y) - \xi(x - y) = \sup_{y \geq x} f(y) - \xi(x - y). \tag{3.1}$$

Then for $u \geq 0$, we have

$$\Phi(f) = \Phi(a) \Rightarrow \Phi(\widehat{f} + u) \geq \Phi(a + u).$$

This statement is weaker than Proposition 3.5. Indeed, fixing $u \geq 0$, we have

$$\widehat{f}(x) + u = \sup_{y} f(y) - \xi(x - y) + u$$

$$\geq \sup\{f(y) - \xi(x - y) + u; \quad \xi(x - y) \leq u\} = \overline{f}(x).$$

Our first task is to show that Proposition 3.1 follows from the following.

Proposition 3.2. *(Basic inequality) Consider a non-decreasing function f from R to R. Then*

$$\left(\int_{\mathsf{R}} e^{f(x)} \, d\mu(x)\right) \left(\int_{\mathsf{R}} e^{-\widehat{f}(x)} \, d\mu(x)\right) \leq 1. \tag{3.2}$$

We collect inequalities for that purpose. We note that

$$\Phi(x) = \frac{1}{2} \int_{-\infty}^{x} e^{-|t|} \, dt = \begin{cases} \frac{1}{2} e^{x} & \text{if } x \leq 0 \\ 1 - \frac{1}{2} e^{-x} & \text{if } x \geq 0. \end{cases}$$

Thus $\Phi(x) = \theta(e^x)$, where $\theta(t) = t/2$ for $t \leq 1$, $\theta(t) = 1 - 1/2t$ for $t \geq 1$. Since $\theta(0) = 0$ and θ' decreases, we have $\theta(t/M) \geq \theta(t)/M$ for $M \geq 1$. Thus

$$v \geq 0 \Rightarrow \Phi(x - v) \geq e^{-v}\Phi(x). \tag{3.3}$$

Observe also that $\Phi(x) \leq \frac{1}{2} e^{x}$.

Lemma 3.3. *Consider a function $g : \mathsf{R} \to \overline{\mathsf{R}}$, such that $\int e^{-g} \, d\mu \leq 1$. Then $\Phi(g + v) \geq \Phi(v)$ for all $v \in \mathsf{R}$.*

Proof: Suppose first that $v \geq 0$. Since

$$1 - \Phi(x) = \Phi(-x) \leq \frac{1}{2} e^{-x},$$

we get

$$\int_{\mathbb{R}} (1 - \Phi(g(x) + v))\, d\mu(x) \le \frac{1}{2}\int e^{-g-v}\, d\mu \le \frac{1}{2} e^{-v} = 1 - \Phi(v)$$

so that

$$\Phi(g + v) = \int \Phi(g(x) + v)\, d\mu(x) \ge \Phi(v).$$

In particular, for $v = 0$, we have

$$\Phi(g) \ge \Phi(0) = \tfrac{1}{2}. \tag{3.4}$$

Suppose now that $v \le 0$. Then, by (3.3) and (3.4), we have

$$\Phi(g + v) = \int \Phi(g(x) + v)\, d\mu(x) \ge e^v \int \Phi(g(x))\, d\mu(x) = e^v \Phi(g) \ge \frac{1}{2} e^v = \Phi(v). \qquad \square$$

We now deduce Proposition 3.1 from Proposition 3.2. We suppose first that $a \le 0$, so that $\Phi(a) = \frac{1}{2}e^a$. Since $\Phi(f(x)) \le \frac{1}{2}e^{f(x)}$, and $\Phi(a) = \Phi(f)$, we get $\int e^f\, d\mu \ge e^a$, so that $\int e^{f-a}\, d\mu \ge 1$. Obviously, we have $\widehat{f - a} = \widehat{f} - a$. The basic inequality (3.2) implies that $\int e^{-(\widehat{f}-a)}\, d\mu \le 1$. Lemma 3.3, used with $v = a + u$, $g = \widehat{f} - a$ concludes the proof in that case.

We suppose now that $a > 0$, and we set $h(x) = -\widehat{f}(-x) - u$. Thus

$$\widehat{h}(x) = \sup_y h(y) - \xi(x - y) = \sup_y -\widehat{f}(-y) - u - \xi(x - y).$$

Since $\widehat{f}(-y) \ge f(-x) - \xi(x - y)$, it follows that

$$\widehat{h}(x) + u \le -f(-x). \tag{3.5}$$

Define b by $\Phi(b) = \Phi(h)$, so that

$$1 - \Phi(b) = \int (1 - \Phi(-\widehat{f}(-x) - u))\, d\mu(x) = \int \Phi(\widehat{f}(-x) + u)\, d\mu(x)$$

$$= \int \Phi(\widehat{f}(x) + u)\, d\mu(x) = \Phi(\widehat{f} + u). \tag{3.6}$$

Since $a \ge 0$, we have, since $f \le \widehat{f} + u$,

$$\tfrac{1}{2} \le \Phi(a) = \Phi(f) \le \Phi(\widehat{f} + u) = 1 - \Phi(b)$$

so that $b \le 0$. Since $b \le 0$, we already know that $\Phi(b + u) \le \Phi(\widehat{h} + u)$. By (3.5) we have, setting $\widetilde{f}(x) = -f(-x)$

$$\Phi(b + u) \le \Phi(\widehat{h} + u) \le \Phi(\widehat{\widetilde{f}}) = 1 - \Phi(f) = 1 - \Phi(a) = \Phi(-a)$$

and thus $a + b + u \le 0$, so that, by (3.6)

$$\Phi(a + u) \le \Phi(-b) = 1 - \Phi(b) = \Phi(\widehat{f} + u). \qquad \square$$

A main step in the proof of the basic inequality is as follows.

Proposition 3.4. *There exists a universal constant K_1 with the following property. Considering a non-decreasing function f from $\mathbb{R}$ to $\mathbb{R}$, such that $\int e^f \, d\mu = 1$. Suppose that the (right) derivative $f'(x)$ of f is $\leq 1/K_1$ for each t. Then we have*

$$\int e^{-f} \, d\mu - K_1 \int e^{-f} f'^2 \, d\mu \leq 1.$$

Lemma 3.5. $\int (f - f(0))^2 \, d\mu \leq 4 \int f'^2 \, d\mu.$

Proof: Consider a function g on $\mathbb{R}$, such that $g(0) = 0$, $\|g'\|_\infty < \infty$. Integrating by parts yields

$$\int_0^\infty g^2(x) e^{-x} \, dx = \int_0^\infty 2g(x) g'(x) e^{-x} \, dx$$

$$\leq 2 \left(\int_0^\infty g(x)^2 e^{-x} \, dx \right)^{1/2} \left(\int_0^\infty g'(x)^2 e^{-x} \, dx \right)^{1/2}$$

by Cauchy-Schwartz. Thus

$$\int_0^\infty g^2(x) e^{-x} \, dx \leq 4 \int_0^\infty g'(x)^2 e^{-x} \, dx.$$

Changing x in $-x$ we get

$$\int_{-\infty}^0 g^2(x) e^{x} \, dx \leq 4 \int_{-\infty}^0 g'(x)^2 e^{x} \, dx$$

so that

$$\int g^2 \, d\mu \leq 4 \int g'^2 \, d\mu. \qquad \square$$

Lemma 3.6. *If $a \geq 0$, $b \leq 0$, we have*

$$\int_a^\infty e^f \, d\mu \leq \frac{K_1}{K_1 - 1} \frac{e^{-a}}{2} e^{f(a)} \tag{3.7}$$

$$\int_{-\infty}^b e^f \, d\mu \geq \frac{K_1}{K_1 + 1} \frac{e^b}{2} e^{f(b)} \tag{3.8}$$

$$\int_{-\infty}^b e^{-f} \, d\mu \leq \frac{K_1}{K_1 - 1} \frac{e^b}{2} e^{-f(b)}. \tag{3.9}$$

Proof: We prove only (3.7), the others being similar. Since $\|f'\|_\infty \leq 1/K_1$, we have, for $x \geq a$

$$f(x) \leq f(a) + (x - a)/K_1$$

so that

$$\int_a^\infty e^{f(x)} \, d\mu(x) \leq \frac{1}{2} \int_a^\infty e^{f(a) + (x-a)/K_1} e^{-x} \, dx$$

$$= \frac{1}{2} e^{f(a) - a/K_1} \int_a^\infty e^{-x(1 - 1/K_1)} \, dx$$

$$= \frac{K_1}{K_1 - 1} \frac{e^{-a}}{2} e^{f(a)}. \qquad \square$$

Lemma 3.7. $|f(0)| \leq 1/2$ if K_1 is large enough.

Proof: Using (3.8), we have

$$1 = \int e^f \, d\mu = \int_{-\infty}^0 e^{f(x)} \, d\mu(x) + \int_0^\infty e^{f(x)} \, d\mu(x)$$

$$\geq \frac{K_1}{2(K_1+1)} e^{f(0)} + \frac{1}{2} e^{f(0)} = \frac{2K_1+1}{2K_1+2} e^{f(0)}$$

so that $e^{f(0)} \leq \frac{2K_1+2}{2K_1+1}$. In a similar way, using (3.7), we get $e^{-f(0)} \geq \frac{2K_1-2}{2K_1-1}$.

We now set

$$c = \inf\{x \in \mathbb{R}, \quad f(x) \geq -1\}; \quad d = \sup\{x \in \mathbb{R}; \quad f(x) \leq 1\}.$$

Observe that, by the previous lemma, we have $c \leq 0 \leq d$.

Lemma 3.8. $K \int_c^d f'^2 \, d\mu \geq e^c + e^{-d} + \int_c^d (f(x) - f(0))^2 \, d\mu(x)$.

Proof: We apply Lemma 3.5 to the function

$$g(x) = \max(-1, \min(1, f(x))) - f(0),$$

and we observe, that, since $|f(0)| \leq 1/2$ by Lemma 3.7, we have $g(x)^2 \geq 1/4$ for $x < d$ or $x > c$.

Lemma 3.9. $f(0)^2 \leq K \int_c^d f'^2 \, d\mu$.

Proof: For $c \leq x \leq d$, we have $-1 \leq f(x) \leq 1$, so that

$$|e^{f(x)} - e^{f(0)}| \leq e^2 |f(x) - f(0)|.$$

Thus, using Cauchy Schwartz and Lemma 3.8

$$\left| \int_c^d (e^{f(x)} - e^{f(0)}) \, d\mu \right| \leq e^2 \int_c^d |f(x) - f(0)| \, d\mu(x) \tag{3.10}$$

$$\leq e^2 \left(\int_c^d |f(x) - f(0)|^2 \, d\mu(x) \right)^{1/2}$$

$$\leq KT$$

where $T^2 = \int_c^d f'^2 \, d\mu$. Now by (3.7), (3.9) and Lemma 3.8 we get

$$1 - \int_c^d e^f \, d\mu = \int_{-\infty}^c e^f \, d\mu + \int_d^\infty e^f \, d\mu$$

$$\leq K(e^c + e^{-d}) \leq KT^2.$$

Since

$$e^{f(0)} - \int_c^d e^{f(0)} \, d\mu \le K(e^c + e^{-d}) \le KT^2,$$

we get

$$|1 - e^{f(0)}| \le K(T + T^2) \le KT$$

since $T \le 1/K_1$. This implies the result. $\qquad\square$

Proof of Proposition 3.4. For $|t| \le 1$, we have $e^{-t} + e^t \le 2 + 4t^2$ (since the function $t^{-2}(e^{-t} + e^t - 2)$ increases). Thus, for $c \le x \le d$ we have

$$e^{-f(x)} \le 2 - e^{f(x)} + 4f^2(x)$$

so that

$$\int_c^d e^{-f} \, d\mu \le \int_c^d (2 - e^f) \, d\mu + 4 \int_c^d f^2 \, d\mu. \tag{3.11}$$

Now

$$\int e^{-f} \, d\mu \le \int_c^d e^{-f} \, d\mu + \int_{-\infty}^c e^{-f} \, d\mu + \int_d^\infty e^{-f} \, d\mu \tag{3.12}$$

$$\le \int_c^d e^{-f} \, d\mu + K(e^c + e^{-d})$$

by (3.9). Since $\int (2 - e^f) \, d\mu = 1$, we have, by (3.7)

$$\int_c^d (2 - e^f) \, d\mu \le 1 - \int_{-\infty}^c (2 - e^f) \, d\mu - \int_d^\infty (2 - e^f) \, d\mu$$

$$\le 1 + K(e^c + e^{-d}).$$

Combining with (3.11), (3.12) yields

$$\int e^{-f} \, d\mu \le 1 + K(e^c + e^{-d}) + 4 \int_c^d f^2 \, d\mu$$

$$\le 1 + K(e^c + e^{-d}) + 8f(0)^2 + 8 \int_c^d (f(x) - f(0))^2 \, d\mu(x)$$

since $f^2(x) \le 2(f(0)^2 + (f(x) - f(0))^2)$. It follows from Lemmas 3.8 and 3.9 that

$$\int e^{-f} \, d\mu \le 1 + K \int_c^d f'^2 \, d\mu$$

$$\le 1 + K \int_c^d f'^2 e^{-f} \, d\mu$$

since $e^{-f(x)} \ge 1/e$ for $x \le d$. $\qquad\square$

Proof of the basic inequality. Step 1. We show that it suffices to consider the case where f is of the type $f(x) = \inf_y g(y) + \xi(x - y)$, where g is a non-decreasing step function (i.e., takes only finitely many values in $\overline{\mathsf{R}}$). Indeed, consider a non-decreasing step function $g \geq \widehat{f}$. Define $f'(x) = \inf_y g(y) + \xi(x - y)$. It is simple to see that $f' \geq f$, and $\widehat{f'} \leq g$. Since $f' \geq f$, we have $\int e^{f'} d\mu \geq 1$. If we know that the result holds for f', we conclude that $\int e^{-\widehat{f'}} d\mu \leq 1$, so that $\int e^{-g} d\mu \leq 1$. Since g is arbitrary, we get $\int e^{-\widehat{f}} d\mu \leq 1$.

Step 2. For $\sigma, \tau \in \mathsf{R}$, we define

$$R(\sigma, \tau) = \{(x, y) ; \quad x \leq \sigma, \quad y \leq \tau \quad \text{or} \quad x \geq \sigma, \quad y \leq \tau + \xi(x - \sigma)\}.$$

It follows from Step 1 that we can assume that the subgraph of f is the intersection of a finite sequence of sets $R(\sigma_i, \tau_i)$, where $\sigma_i < \sigma_{i+1}$, $\tau_i < \tau_{i+1}$.

For $s \leq x$, we define

$$H(x, s) = \sup_{s \leq y \leq x} f(y) - \xi(y - s).$$

That $H(x, x) = f(x)$. We now prove the following fact. Fix x, and consider the largest i such that $(x, f(x))$ is on the boundary of $R(\sigma_i, \tau_i)$. Assume that $\sigma_i < x$. Then for $\sigma_i < s < x$, we have $H(x, s) = f(x) - \xi(x - s)$.

Consider $s \leq y < x$. We have $f(y) \leq \tau_i + \xi(y - \sigma_i)$, so that

$$f(y) - \xi(y - s) \leq \tau_i + \xi(y - \sigma_i) - \xi(y - s).$$

Since $\sigma_i \leq s$, and since ξ' increases on R^+, the function $z \to \xi(z - \sigma_i) - \xi(z - s)$ increases. Thus

$$\tau_i + \xi(y - \sigma_i) - \xi(y - s) \leq \tau_i + \xi(x - \sigma_i) - \xi(x - s) = f(x) - \xi(x - s).$$

Step 3. We define

$$W(x) = \int_{-\infty}^{x} e^{-H(x,s)} d\mu(s).$$

We note that $H(x, s) \leq \widehat{f}(s)$.

Thus, for $y < x$

$$W(x) \geq \int_{-\infty}^{y} e^{-H(x,s)} d\mu(s) \geq \int_{-\infty}^{y} e^{-\widehat{f}(s)} d\mu(s),$$

so that

$$\lim_{x \to \infty} W(x) \geq \int e^{-\widehat{f}} d\mu.$$

To conclude, it suffices to show that $\int W'(x)\,dx \leq 1$. Since $H(x,x) = f(x)$, we have

$$W'(x) = e^{-f(x)}\varphi(x) - \int_{-\infty}^{x} \frac{\partial}{\partial x}H(x,s)e^{-H(x,s)}\,d\mu(s)\,.$$

(Thus we can differentiate under the integral sign follows from the fact that $\frac{\partial}{\partial x}H(x,s)$ is bounded by $\|\xi'\|_{\infty}$.) Since $H(x,s) \leq f(x)$, we have

$$W'(x) \leq e^{-f(x)}\left(\varphi(x) - \int_{-\infty}^{x} \frac{\partial}{\partial x}H(x,s)\,d\mu(s)\right)\,. \tag{3.13}$$

Consider the largest i such that x is on the boundary of $R(\sigma_i, \tau_i)$. Suppose first that $\sigma_i < x$. Step 2 shows that for $\sigma_i < s < x$, we have $H(x,s) = f(x) - \xi(x-s)$ so that

$$\frac{\partial}{\partial x}H(x,s) = f'(x) - \xi'(x-s)$$

$$= \xi'(x-\sigma_i) - \xi'(x-s)$$

since $f(y) = \tau_i + \xi(y - \sigma_i)$ for y close enough to x.

We recall that $\xi'(t) = \frac{t}{L(t+1)}$. If $x \geq \sigma_i + 2$, we thus have, for $s \geq x - 1$

$$\frac{\partial}{\partial x}H(x,s) \geq \frac{2}{3L} - \frac{1}{2L} = \frac{1}{6L}\,.$$

If $x \leq \sigma_i + 2$, for $s \geq (x + \sigma_i)/2$, setting $t = x - \sigma_i \leq 2$, we have

$$\frac{\partial}{\partial x}H(x,s) \geq \frac{1}{L}\left(\frac{t}{1+t} - \frac{t/2}{1+t/2}\right) = \frac{1}{L}\frac{t}{2(t+1)(t+2)} \geq \frac{t}{24L}\,.$$

Thus, in both cases $\frac{\partial}{\partial x}H(x,s)$ is $\geq \frac{1}{KL}\min(1, x-\sigma_i)$ on an interval of length $\geq \frac{1}{2}\min(2, x-\sigma_i)$. Thus, since $\varphi(y) \geq \frac{1}{e^2}\varphi(x)$ for $|y - x| \leq 2$, and since $\frac{\partial}{\partial x}H(x,s) \geq 0$, we have

$$\int_{-\infty}^{x} \frac{\partial}{\partial x}H(x,s)\,d\mu(s) \geq \frac{1}{KL}\varphi(x)\min(1, (x-\sigma_i)^2)$$

$$\geq \frac{1}{KL}\varphi(x)\frac{(x-\sigma_i)^2}{(1+|x-\sigma_i|)^2}$$

$$= \frac{1}{KL}\varphi(x)(L\xi'(x-\sigma_i))^2$$

$$= \frac{L}{K}\varphi(x)\xi'(x-\sigma_i)^2$$

$$= \frac{L}{K}\varphi(x)f'(x)^2\,.$$

Combining with (3.13) we get

$$W'(x) \leq e^{-f(x)}\varphi(x)\left(1 - \frac{L}{K}f'(x)^2\right)\,. \tag{3.14}$$

Suppose now that $x \leq \sigma_i$. In that case $f'(x) = 0$, so that (3.14) still hold. If we take $L = KK_1$, where K_1 is the constant of Proposition 3.6, we see from (3.14) that $\int_{\mathbf{R}} W'(x)dx \leq 1$. This completes the proof.

4. Variational arguments

We will now show that, provided L is large enough, Proposition 2.5 follows from Proposition 3.1. The method is as follows. We fix the values of a and u. We show that there exists an f such that $\Phi(f) = \Phi(a)$, and that $\Phi(\overline{f})$ is as small as possible. We then show that $\overline{f}$ being given by (2.3), we have $\overline{f} = \widehat{f} + u$ (in which case $\Phi(\overline{f}) \geq \Phi(a+u)$ by Proposition 3.1).

The first task is to show that the infimum of $\Phi(\overline{f})$ is actually obtained. We recall that a, u are fixed, and we set

$$\mathcal{F} = \{f : \mathbb{R} \to \overline{\mathbb{R}}, \quad f \text{ non-decreasing}, \quad \Phi(f) = \Phi(a)\}.$$

Lemma 4.1. *Consider an ultrafilter $\mathcal{U}$ on $\mathcal{F}$. Define $f(x) = \lim\limits_{g \to \mathcal{U}} g(x)$. Then $f \in \mathcal{F}$.*

Proof: It is a well known (and elementary) fact that the map $h \to \int_0^1 h(x)\,dx$ is pointwise continuous on the set $\mathcal{G}$ of non-decreasing functions from $[0,1]$ to $[0,1]$. The map $f \to \Phi(f)$ is thus pointwise continuous on the set of $\mathcal{G}'$ all non-decreasing maps from $\overline{\mathbb{R}}$ to $\overline{\mathbb{R}}$, as is seen by transporting $\mathcal{G}'$ to $\mathcal{G}$ by the map $f \to \Phi \circ f \circ \Phi^{-1}$. $\qquad\square$

We define $\sigma > 0$ by $\xi(\sigma) = u$. Since $\xi(\sigma) \leq \sigma/L$, we have $\sigma \geq Lu$. We can and do assume $L \geq 2$, so that $\sigma \geq 2u$.

Lemma 4.2. *Consider an ultrafilter $\mathcal{U}$ on $\mathcal{F}$. Then if $f \in \mathcal{F}$ is given by $f(x) = \lim\limits_{g \to \mathcal{U}} g(x)$, we have $\lim\limits_{g \to \mathcal{U}} \Phi(\overline{g}) \geq \Phi(\overline{f})$.*

Proof: Since $\overline{g}$ is non-decreasing whenever g is, and since (as mentioned in the proof of Lemma 4.1) Φ is pointwise continuous on the set of non-decreasing functions, it suffices to show that for all x, $\lim\limits_{g \to \mathcal{U}} \overline{g}(x) \geq \overline{f}(x)$. Given y with $|y - x| \leq \sigma$, we have

$$g(y) + u - \xi(y - x) \leq \overline{g}(x).$$

Taking the limit along $\mathcal{U}$ gives

$$f(y) + u - \xi(y - x) \leq \lim_{g \to \mathcal{U}} \overline{g}(x)$$

and thus $\overline{f}(x) \leq \lim\limits_{g \to \mathcal{U}} \overline{g}(x)$ by definition of $\overline{f}$. $\qquad\square$

Lemmas 4.1, 4.2 implies that we can find $f \in \mathcal{F}$ for which $\Phi(\overline{f})$ is minimal. Observe that $\Phi(\overline{f}) \leq \Phi(a+u)$, since the constant function a belongs to $\mathcal{F}$. We fix such an f, and we want to prove that $\overline{f} = \widehat{f} + u$.

Lemma 4.3. *We have $f(2u) \in \mathbb{R}$, $f(\sigma - 2u) \in \mathbb{R}$.*

Proof: We observe that $\overline{f}(x) \geq f(x + \sigma) - \xi(\sigma) + u = f(x + \sigma)$. Set $f'(x) = f(x + \sigma)$. It suffice to show that if either $f(2u) = -\infty$ or $f(\sigma - 2u) = \infty$ we would have $\Phi(f') > \Phi(a + u)$.

Consider the function g from $\mathbb{R}$ to $\overline{\mathbb{R}}$ given by

$$g(y) = \inf\{x; \quad f(-x) \leq y\}.$$

It is simple to see that $\Phi(g) = \Phi(f)$, since, by Fubini theorem, both numbers are equal to

$$\mu^2(\{(x,y); \quad y \leq f(x)\}).$$

We set

$$g'(y) = \inf\{x; \quad f'(-x) \leq y\}.$$

Thus $g' = g + \sigma$, and $\Phi(g') = \Phi(f')$. Thus $\Phi(f') = \Phi(g + \sigma)$. Thus it suffices to show that if either $g \leq -2u$ (case $f(2u) = -\infty$) or $g \geq -\sigma + 2u$ (case $f(\sigma - 2u) = \infty$) we have $\Phi(g + \sigma) > \Phi(a + u)$.

Case $g \leq -2u$. Since $\Phi(g) = \Phi(a)$, we have $a \leq -2u$. We have

$$\Phi(g + \sigma) \geq \Phi(g + 2u) = \int \Phi(g(x) + 2u)\, d\mu(x)$$

$$= e^{2u} \int \Phi(g(x))\, d\mu(x)$$

$$= e^{2u}\Phi(a) = \Phi(a + 2u) > \Phi(a + u)$$

since $\Phi(t + 2u) = e^{2u}\Phi(t)$ for $t + 2u \leq 0$, and since $a \leq -2u$.

Case $g \geq -\sigma + 2u$. We define b by $\Phi(g + \sigma) = 1 - \Phi(b)$, so that $b \leq -2u$. Let $h = -(g + \sigma)$, so that $\Phi(h) = \Phi(b)$. Since $h \leq -2u$, the preceding case shows that $\Phi(h + 2u) \geq \Phi(b + 2u)$. But

$$\Phi(h + 2u) = 1 - \Phi(g + \sigma - 2u) \leq 1 - \Phi(g) = 1 - \Phi(a) = \Phi(-a).$$

Thus $-a \geq b + 2u$, so that $-b \geq a + 2u$, and

$$\Phi(g + \sigma) = 1 - \Phi(b) = \Phi(-b) \geq \Phi(a + 2u) > \Phi(a + u). \qquad \square$$

We denote by $I =]\alpha, \beta[$ the interior of $f^{-1}(\mathbb{R})$. Thus $\alpha \leq 2u$, $\beta \geq \sigma - 2u$. We set $J = I - \sigma$. We note that

$$x < \alpha - \sigma \Rightarrow \overline{f}(x) = -\infty; \qquad x > \beta - \sigma \Rightarrow \overline{f}(x) = +\infty$$

$$x \in J \Rightarrow \overline{f}(x) \in \mathbb{R}.$$

We now start to exploit the fact that $\Phi(\overline{f})$ is as small as possible. For the convenience of notations, we write $(f + g)^-$ for $\overline{f + g}$.

Lemma 4.4. *Consider a bounded continuous function v from $\mathbb{R}$ to $\mathbb{R}$, that is zero outside I. Suppose that $\int v\varphi(f)\,d\mu \geq 0$ (resp. > 0). Then*

$$\limsup_{s \to 0+} s^{-1}(\Phi((f + sv)^-) - \Phi(\bar{f})) \geq 0 \qquad (resp. > 0).$$

Proof: Since $\varphi \leq 1/2$, Φ is lipschitz. Since v is bounded, by dominated convergence we have

$$\int v\varphi(f)\,d\mu = \lim_{s \to 0} s^{-1}(\Phi(f + sv) - \Phi(f)).$$

Set

$$\delta = \int v\varphi(f)\,d\mu; \quad \gamma = \limsup_{s \to 0+} s^{-1}(\Phi((f + sv)^-) - \Phi(\bar{f})).$$

and fix $\delta' < \delta$, $\gamma' > \gamma$. Thus for all s small enough we have

$$\Phi(f + sv) \geq \Phi(f) + \delta's \tag{4.1}$$

$$\Phi((f + sv)^-) \leq \Phi(\bar{f}) + \gamma's. \tag{4.2}$$

Consider the non-decreasing rearrangement g_s of $f + sv$, defined in Proposition 2.6. From that proposition follows that $\Phi(g_s) = \Phi(f + sv)$, $\Phi(\bar{g}_s) \leq \Phi((f + sv)^-)$. Thus, from (4.2), for s small enough, we have

$$\Phi(\bar{g}_s) \leq \Phi(\bar{f}) + s\gamma'. \tag{4.3}$$

Consider the number $t(s)$ such that $\Phi(g_s + t(s)) = \Phi(a)$, so that $g_s + t(s) \in \mathcal{F}$. Since $\Phi(\bar{f})$ is the minimum of $\Phi(\bar{g})$ for $g \in \mathcal{F}$, we have

$$\Phi(\bar{f}) \leq \Phi((g_s + t(s))^-) = \Phi(\bar{g}_s + t(s)).$$

Combining with (4.3), we see that

$$\Phi(\bar{g}_s + t(s)) - \Phi(\bar{g}_s) \geq -s\gamma'. \tag{4.4}$$

It is clear that $g_s + t(s)$ is the non-increasing rearrangement of $f + sv + t(s)$. Thus

$$\Phi(f) = \Phi(a) = \Phi(g_s + t(s)) = \Phi(f + sv + t(s)),$$

and by (4.1), for s small enough, we have

$$\Phi(f + sv) - \Phi(f + sv + t(s)) \geq \delta's. \tag{4.5}$$

Since $\Phi(x) - \Phi(y) \le x - y$ for $x > y$, we have, for all functions h

$$\forall t \ge 0, \qquad \Phi(h+t) - \Phi(h) \le t. \tag{4.6}$$

We suppose now that $\delta' > 0$. Since Φ is increasing, (4.5) show that $t(s) \le 0$, and using (4.6) for $t = -t(s)$, $h = f + sv + t(s)$, we get $-t(s) \ge \delta's$, i.e. $t(s) \le -\delta's$. By (4.4) we now have

$$\gamma' \ge \int s^{-1}(\Phi(\overline{g}_s(x)) - \Phi(\overline{g}_s(x) + t(s)))\, d\mu(x)$$

$$\ge \int s^{-1}(\Phi(\overline{g}_s(x)) - \Phi(\overline{g}_s(x) - \delta's))\, d\mu(x).$$

For $A > 0$, $x, y \in [-A, A]$, $x < y$, we have

$$\Phi(y) - \Phi(x) \ge (x - y)\varphi(A).$$

It clearly follows that for some constant B that depends on f, u only, we have $\gamma' \ge B\delta'$. Thus $\delta > 0$ implies $\gamma > 0$.

Suppose now that $\gamma' < 0$. By (4.4) we see now that $t(s) \ge 0$, and by (4.6) that $t(s) \ge (-\gamma')s$. By (4.5) we have

$$s^{-1}(\Phi(f + sv + (-\gamma')s) - \Phi(f + sv)) \le -\delta'.$$

Letting $s \to 0$, we get by Fatou's lemma that

$$(-\gamma') \int \varphi(f)\, d\mu \le -\delta'$$

so that $\gamma < 0$ implies $\delta < 0$, and thus $\delta \ge 0$ implies $\gamma \ge 0$. $\qquad\square$

For $x \in J$, we define

$$A_x = \bigcap_{\varepsilon > 0} \text{closure } \{y;\quad x \le y \le x + \sigma,\quad \overline{f}(x) \le f(y) - \xi(y - x) + u + \varepsilon\}.$$

While the idea of the following result is well known, we provide a proof since we cannot find an exact reference.

Lemma 4.5. $\displaystyle \lim_{s \to 0_+} s^{-1}((f + sv)^-(x) - \overline{f}(x)) = \sup_{z \in A_x} v(z).$

Proof: a) Consider $z \in A_x$, and $\varepsilon > 0$. Thus we can find y such that $x \le y \le x + \sigma$ and that

$$v(y) \ge v(z) - \varepsilon;\qquad \overline{f}(x) \le f(y) - \xi(y - x) + u + \varepsilon.$$

Since

$$(f + sv)^-(x) \geq f(y) + sv(y) - \xi(y - x) + u\,,$$

we get

$$(f + sv)^-(x) - \overline{f}(x) \geq sv(y) - \varepsilon \geq sv(z) - \varepsilon(1 + s)\,.$$

This holds for all $\varepsilon > 0$. Thus

$$(f + sv)^-(x) - \overline{f}(x) \geq sv(z)$$

and thus

$$\liminf_{s \to 0_+} s^{-1}((f + sv)^-(x) - \overline{f}(x)) \geq v(z)\,.$$

Since z is arbitrary in A_ε, we have

$$\liminf_{s \to 0_+} s^{-1}((f + sv)^-(x) - \overline{f}(x)) \geq \sup_{z \in A_x} v(z)\,.$$

b) Consider now $B > \sup\limits_{z \in A_x} v(z)$. Since v is continuous, the set $\{v < B\}$ is a neighborhood of A_x. Thus, by definition of A_x we can find $\varepsilon > 0$ such that

$$f(y) - \xi(y - x) + u + \varepsilon \geq \overline{f}(x) \Rightarrow v(y) < B\,.$$

Distinguishing whether $v(y) < B$ or not, we have

$$(f + sv)^-(x) = \sup\{f(y) - \xi(y - x) + u + sv(y);\quad x \leq y \leq x + \sigma\}$$
$$\leq \max\{\overline{f}(x) + sB,\quad \overline{f}(x) - \varepsilon + s\|v\|_\infty\}$$

so that

$$s^{-1}((f + sv)^-(x) - \overline{f}(x)) \leq \max\left(B, \|v\|_\infty - \frac{\varepsilon}{s}\right)$$

and hence

$$\limsup_{s \to 0_+} s^{-1}((f + sv)^-(x) - \overline{f}(x)) \leq B$$

for all $B > \sup\limits_{z \in A_x} v(z)$. $\qquad\qquad\qquad\qquad\qquad\qquad\qquad\qquad\square$

Lemma 4.6. *Consider* $x_1, x_2 \in J$, $x_1 < x_2$, $z_1 \in A_{x_1}$, $z_2 \in A_{x_2}$. *Then* $z_1 \leq z_2$.

Proof: We first observe that for $z \in A_x$, there is a sequence (z_n) converging to z such that

$$\overline{f}(x) = \lim_{n \to \infty} f(z_n) - \xi(z_n - x) + u\,.$$

Since f is non-decreasing, setting $f^+(z) = \lim_{y \to z, y > z} f(y)$, we have

$$\overline{f}(x) \leq f^+(z) - \xi(z - x) + u. \tag{4.7}$$

We now argue by contradiction, and assume that $z_2 < z_1$. Thus $x_1 < x_2 \leq z_2 < z_1 \leq x_1 + \sigma$. For $x_1 \leq y \leq x_1 + \sigma$, we have

$$f(y) - \xi(y - x_1) + u \leq \overline{f}(x_1)$$

so that, letting $y \to z_2$, $y > z_2$ we get

$$f^+(z_2) - \xi(z_2 - x_1) + u \leq \overline{f}(x_1) \leq f^+(z_1) - \xi(z_1 - x_1) + u \tag{4.8}$$

where the second inequality follows from (4.7). A similar argument, using now the fact that $z_1 \leq x_1 + \sigma < x_2 + \sigma$ gives

$$f^+(z_1) - \xi(z_1 - x_2) + u \leq \overline{f}(x_2) \leq f^+(z_2) - \xi(z_2 - x_2) + u.$$

Adding with (4.8) gives

$$\xi(z_1 - x_1) + \xi(z_2 - x_2) \leq \xi(z_2 - x_1) + \xi(z_1 - x_2).$$

so that

$$\xi(z_1 - x_2) - \xi(z_1 - x_1) \geq \xi(z_2 - x_2) - \xi(z_2 - x_1).$$

But since ξ' increases strictly on $\mathbf{R}^+$, and since $x_1 < x_2$, the function $z \to \xi(z - x_2) - \xi(z - x_1)$ decreases strictly for $z > x_1$. This contradicts the fact that $z_1 > z_2$. $\square$

Proposition 4.7. *For all $x \in J$, A_x has exactly one point.*

Proof: By definition, A_x is not empty. Suppose that for some $x \in J$, A_x contains two points. Then we can find $z_1 < z_2$ such that A_x contains a point $< z_1$ and a point $> z_2$. By Lemma 4.6, for $y \neq x$, we have $A_y \cap [z_1, z_2] = \emptyset$. Consider now a positive continuous function v that is supported by $[z_1, z_2]$. It follows from Lemma 4.5 that for $y \neq x$, we have

$$\lim_{s \to 0_+} s^{-1}\big((f + sv)^-(y) - \overline{f}(y)\big) = 0.$$

By dominated convergence, we have

$$\lim_{s \to 0_+} s^{-1}\big(\Phi((f + sv)^-) - \Phi(\overline{f})\big) = 0.$$

But since $\int v\varphi(f)\,d\mu > 0$, this contradicts Lemma 4.3. □

We define $A(x)$ by $A_x = \{A(x)\}$. We note that, by (4.7) we have

$$\overline{f}(x) \leq f^+(A(x)) - \xi(A(x) - x) + u. \tag{4.9}$$

Moreover, if f is continuous at $A(x)$, we have

$$f(x) = f(A(x)) - \xi(A(x) - x) + u. \tag{4.9'}$$

For $y \in I$, we set

$$w(y) = \sup\{x \in J; \quad A(x) \leq y\}.$$

Proposition 4.8. *For some constant C we have*

$$\forall y \in I, \quad \int_{\alpha-\sigma}^{w(y)} \varphi(\overline{f}(x))\,d\mu(x) = C \int_{\sigma}^{y} \varphi(f(x))\,d\mu(x). \tag{4.10}$$

In particular w is one to one and continuous.

Proof: It follows from Lemma 4.5 and dominated convergence that, for a bounded continuous function v supported by J, we have

$$\lim_{s \to 0_+} s^{-1}(\Phi((f + sv)^-) - \Phi(\overline{f})) = \int_J v(A(x))\varphi(\overline{f}(x))\,d\mu(x).$$

By Lemma 4.3, we have

$$\int_I v(x)\varphi(f(x))\,d\mu(x) = 0 \Rightarrow \int_J v(A(x))\varphi(\overline{f}(x))\,d\mu(x) = 0.$$

Thus, there exists a constant C such that for all continuous bounded functions v with support in I, we have

$$\int_J v(A(x))\varphi(\overline{f}(x))\,d\mu(x) = C \int_I v(x)\varphi(f(x))\,d\mu(x).$$

If we approximate the indicator function of $]\alpha, y]$ by v we get

$$\int_B \varphi(\overline{f}(x))\,d\mu(x) = C \int_\alpha^y \varphi(f(x))\,d\mu(x)$$

where $B = \{x \geq \alpha - \sigma, \quad A(x) \leq y\}$.

We observe that $A(w(x)) = x$ for $x \in I$ and $w(A(x)) = x$ for $x \in J$.

We now define

$$Z = \{y \in I; \quad w(y) = y - \sigma\}.$$

Since w is continuous, Z is closed in I. Our objective is to prove that $Z = \emptyset$ (at which point the proof will be almost finished).

Lemma 4.9. *Suppose that f is not continuous at $y \in I$. Then $y \in Z$.*

Proof: Consider $z < y$. By (4.9) we have

$$\bar{f}(w(z)) \leq f^+(A(w(z))) - \xi(A(w(z)) - w(z)) + u$$

$$\leq f^+(z) - \xi(z - w(z)) + u$$

$$< f^+(y) - \xi(y - w(z)) + u$$

as soon as $(y - z)\|\xi'\|_\infty \leq \frac{1}{L}(y - z) < f^+(y) - f^-(y)$. Thus, by definition of $\bar{f}$, we must have $w(z) + \sigma \leq y$. Since w is continuous, letting $z \to y$ we have $w(y) + \sigma \leq y$, so that $w(y) = y - \sigma$. $\square$

Lemma 4.10. *For $x \in I$, we have $f(x + \sigma) \leq \bar{f}(x) \leq f(x + \sigma) + u$. Thus*

$$e^{-u}\varphi(f(x + \sigma)) \leq \varphi(\bar{f}(x)) \leq e^u \varphi(f(x + \sigma)). \tag{4.11}$$

Proof: We have

$$\bar{f}(x) \geq f(x + \sigma) - \xi(\sigma) + u = f(x + \sigma)$$

and also

$$\bar{f}(x) \leq \sup_{|y - x| \leq \sigma} f(y) - \xi(y - x) + u \leq f(x + \sigma) + u. \qquad \square$$

Lemma 4.11. *Consider $y_1 < y_2$, $y_1, y_2 \in Z \cup \{\alpha, \beta\}$. Then if we set $-|-\infty - \sigma| + |-\infty| = -\sigma$; $-|\infty - \sigma| + |\infty| = \sigma$, we have*

$$e^{-u - |y_1 - \sigma| + |y_1|} \leq C \leq e^{u - |y_2 - \sigma| + |y_2|}.$$

Proof: By (4.10) we have

$$\int_{y_1 - \sigma}^{y_2 - \sigma} \varphi(\bar{f}(x)) \, d\mu(x) = C \int_{y_1}^{y_2} \varphi(f(x)) \, d\mu(x).$$

Thus, by (4.11),

$$C \int_{y_1}^{y_2} \varphi(f(x)) \, d\mu(x) \leq e^u \int_{y_1 - \sigma}^{y_2 - \sigma} \varphi(f(x + \sigma))\varphi(x) \, dx$$

$$\leq e^u \int_{y_1}^{y_2} \varphi(f(x))\varphi(x - \sigma) \, dx$$

$$\leq e^u \sup_{y_1 \leq x \leq y_2} \frac{\varphi(x - \sigma)}{\varphi(x)} \int_{y_1}^{y_2} \varphi(f(x)) \, d\mu(x).$$

Thus

$$C \leq e^u \sup\{e^{-|x - \sigma| + |x|}; \quad y_1 \leq x \leq y_2\}$$

$$\leq e^{u - |y_2 - \sigma| + |y_2|}$$

since the function $x \to -|x - \sigma| + |x|$ is non-decreasing. The other inequality is proved in as similar way. $\square$

Corollary 4.12. *For $y_1, y_2 \in Z$, we have*

$$-|y_1 - \sigma| + |y_1| \le 2u + (-|y_2 - \sigma| + |y_2|).$$

(Note that we no longer assume $y_1 \le y_2$. The case of interest is actually $y_1 > y_2$.)

Proof: We use Lemma 4.11 for the pair α, y_2, and then for the pair y_1, β. □

Lemma 4.13. *If $y \in I$, $y \notin Z$, we have*

$$\limsup_{z \to y^+} \frac{f(z) - f(y)}{z - y} \le \xi'(\sigma).$$

Proof: By Lemma 4.9, f is continuous at y. Thus for $|z - w(y)| < \sigma$, we have

$$f(z) - \xi(z - w(y)) + u \le \overline{f}(w(y)) = f(y) - \xi(y - w(y)) + u.$$

Using this for $y < z < w(y) + \sigma$, we get

$$\frac{f(z) - f(y)}{z - y} \le \frac{\xi(z - w(y)) - \xi(y - w(y))}{z - y}.$$

This implies the result since $\xi'(y - w(y)) \le \xi'(\sigma)$ as ξ' increases on $\mathbf{R}^+$. □

Lemma 4.14. $\sigma \xi'(\sigma) \le 2u.$

Proof:

$$u = \xi(\sigma) = \frac{1}{L} \int_0^\sigma \frac{x}{1 + x} \, dx \ge \frac{1}{L(1 + \sigma)} \int_0^\sigma x \, dx$$

$$= \frac{1}{2L} \frac{\sigma^2}{1 + \sigma} = \frac{1}{2} \sigma \xi'(\sigma).$$ □

Lemma 4.15. *Suppose $[x, A(x)] \cap Z = \emptyset$. Then*

$$f(x) + u \le \overline{f}(x) \le f(x) + 3u.$$

Proof: Since $\xi(0) = 0$, we have $\overline{f}(x) \ge f(x) + u$. Since $\overline{f}$ is continuous at $A(x)$ by Lemma 4.9, we have by (4.9′) that

$$\overline{f}(x) = f(A(x)) - \xi(A(x) - x) + u$$

$$\le f(A(x)) + u.$$

By Lemmas 4.13, 4.14, we have

$$f(A(x)) \le f(x) + (A(x) - x)\xi'(\sigma) \le f(x) + \sigma \xi'(\sigma) \le f(x) + 2u.$$ □

Lemma 4.16. *a) Suppose that for some $\alpha' > \alpha$, we have $(\alpha, \alpha') \cap Z = \emptyset$. Then $\alpha = -\infty$ and $C \le e^{3u}$.*

b) Suppose that for some $\beta' < \beta$, we have $(\beta', \beta) \cap Z = \emptyset$. Then $\beta = \infty$ and $C \ge e^{-3u}$.

Proof: We prove only a) since b) is similar. The fact that $\alpha = -\infty$ follows from Lemmas 4.9 and 4.13. By Proposition 4.8 we have

$$C \int_{-\infty}^{\alpha'} \varphi(f(x)) \, d\mu(x) = \int_{-\infty}^{w(\alpha')} \varphi(\overline{f}(x)) \, d\mu(x).$$

Now for $x < w(\alpha')$, we have $A(x) < \alpha'$, so that $|\overline{f}(x) - f(x)| \le 3u$ by Lemma 4.15 and $\varphi(\overline{f}(x)) \le e^{3u} \varphi(f(x))$. Thus

$$C \int_{-\infty}^{\alpha'} \varphi(f(x)) \, d\mu(x) \le e^{3u} \int_{-\infty}^{w(\alpha')} \varphi(f(x)) \, d\mu(x)$$

$$\le e^{3u} \int_{-\infty}^{\alpha'} \varphi(f(x)) \, d\mu(x). \qquad \square$$

Proposition 4.17. *Suppose that $L \ge 30$. Then $Z = \emptyset$.*

Proof: **Step 1** We have $\xi(x) < |x|/L$, so that $\sigma > Lu \ge 30u$. Since $\alpha \le 2u$, $\beta \ge \sigma - 2u$, the value of the function $-|x - \sigma| + |x|$ at β (resp. α) is at least $\sigma - 4u$ (resp. at most $-\sigma + 4u$). Thus Corollary 4.12 shows that α, β cannot be both cluster points of Z, for otherwise, we would have

$$\sigma - 4u \le 2u + (-\sigma + 4u).$$

For definiteness, we assume that $\alpha \notin Z$. By Lemma 4.16 a), we have $\alpha = -\infty$, $C \le e^{3u}$. By Lemma 4.11, we have $-|y - \sigma| + |y| \le 4u$ for $y \in Z$, so that $y \le \sigma/2 + 2u$. Since $\beta \ge \sigma - u$, and $\sigma > 6u$, β cannot be a cluster point of Z. By Proposition 4.15 b), we have $\beta = \infty$, $C \ge e^{-3u}$. Thus, by Lemma 4.11, we have $-4u \le -|y - \sigma| + |y|$ for $y \in Z$, so that $z \ge \sigma/2 - 2u$. Thus $Z \subset [\frac{\sigma}{2} - 2u, \ \frac{\sigma}{2} + 2u]$.

Step 2 Suppose now that $Z \ne \emptyset$. Consider the smallest point $z \in Z$. We have

$$\int_{-\infty}^{z-\sigma} \varphi(\overline{f}(x)) \, d\mu(x) = C \int_{-\infty}^{z} \varphi(f(x)) \, d\mu(x).$$

By Lemma 4.14, we have $\varphi(\overline{f}(x)) \le e^{3u} \varphi(f(x))$ for $x \le z - \sigma$; since $z \ge 0$, we have

$$C \int_{-\infty}^{0} \varphi(f(x)) \, d\mu(x) \le e^{3u} \int_{-\infty}^{z-\sigma} \varphi(f(x)) \, d\mu(x).$$

By Lemma 4.13, for $x \le 0$, we have $|f(x) - f(0)| \le |x|\xi'(\sigma)$, so that

$$\varphi(f(x)) \ge e^{x\xi'(\sigma)}\varphi(f(0)).$$

For $x \le z - \sigma$, we have $|f(x) - f(z - \sigma)| \le |z - \sigma - x|\xi'(\sigma)$, again by Lemma 4.13, so that

$$\varphi(f(x)) \le e^{(z-\sigma-x)\xi'(\sigma)}\varphi(f(z - \sigma)).$$

The proof of Lemma 3.6 then shows that (since $z - \sigma \le 0$)

$$\int_{-\infty}^{0} \varphi(f(x))\, d\mu(x) \ge \frac{1}{2}(1 + \xi'(\sigma))^{-1}\varphi(f(0))$$

$$\int_{-\infty}^{z-\sigma} \varphi(f(x))\, d\mu(x) \le \frac{1}{2}(1 - \xi'(\sigma))^{-1}e^{-|z-\sigma|}\varphi(f(z - \sigma)).$$

Thus

$$C \le e^{3u}(1 + \xi'(\sigma))(1 - \xi'(\sigma))^{-1}e^{-|z-\sigma|}\frac{\varphi(f(z - \sigma))}{\varphi(f(0))}.$$

Now, since $|z - \sigma| \le \sigma$

$$|f(z - \sigma) - f(0)| \le \sigma\xi'(\sigma) \le 2u$$

so that $\varphi(f(z - \sigma)) \le e^{2u}\varphi(f(0))$. Since $|z - \sigma| \ge \sigma/2 - 2u$, we have

$$C \le e^{7u-\sigma/2}\frac{1 + \xi'(\sigma)}{1 - \xi'(\sigma)}.$$

We note that $\frac{1+x}{1-x} \le 1 + 3x$ for $x \le 1/3$. Since $\xi'(\sigma) \le 1/L$, and since $\xi'(\sigma) \le \sigma/L$, we get since $\sigma \ge Lu$

$$C \le e^{7u-\sigma/2+3\xi'(\sigma)} \le e^{7u-\sigma\left(\frac{1}{2}-\frac{3}{L}\right)}$$

$$\le e^{7u-Lu\left(\frac{1}{2}-\frac{3}{L}\right)}.$$

Since $L \ge 30$, this contradicts the fact that $C \ge e^{-3u}$. $\qquad\square$

We can now finish the proof of Theorem 1.1.

Proposition 4.18. *Suppose that $Z = \emptyset$. Then $\overline{f} = \widehat{f} + u$.*

Proof: We have shown that when $Z = \emptyset$, $I = \mathbb{R}$. By Lemma 4.13, for $y \ge x + \sigma$, we have, since ξ' is increasing

$$f(y) - \xi(y - x) \le f(x + \sigma) - \xi(\sigma)$$

so that

$$f(x + \sigma) - \xi(\sigma) + u \ge f(y) - \xi(y - x) + u.$$

Thus

$$\widehat{f}(x) + u = \sup_y f(y) - \xi(y - x) + u \le \sup_{y \le x+\sigma} f(y) - \xi(y - x) + u = \overline{f}(x). \qquad\square$$

5. Proof of Theorem 1.2

For a compact set A of $\mathbb{R}^n$, we set

$$h_A(x) = \inf \left\{ u \geq 0 \; ; \; x \in A + V_n(u) \right\} .$$

To prove Theorem 1.2, it suffices, by (2.0) and an obvious approximation argument, to show that for all n we have

$$\int_{\mathbb{R}^n} \exp h_A(x) d\mu^n(x) \leq 2/\mu^n(A) , \qquad (H_n)$$

provided the parameter L has been chosen large enough.

The proof of (H_n) will be by induction over n. We first consider the case $n = 1$. Define a by $\mu(A) = e^{-a}$. Since $\mu([-a, a]) = 1 - e^{-a}$, there exists $b \in A$ with $|b| \leq a$. Thus $|x - b| \leq |x| + a$ and since $\xi(u) \leq u/L$, we have $h_A(x) \leq (|x| + a)/L$. Thus, if we assume $L \geq 2$, we have

$$\int_{\mathbb{R}} \exp h_A(x) d\mu(x) \leq \tfrac{1}{2} \int_{\mathbb{R}} e^{(|x|+a)/L} e^{-|x|} dx$$

$$= \frac{1}{1 - 1/L} e^{a/L} \leq 2e^a = 2/\mu(A) .$$

This proves (H_1). Let us point out that a more cautious computation using e.g. Proposition 2.2 allows the removal of the factor 2 in the right hand side of (H_1) (and, as the proof will show, also in the right hand side of Theorem 1.2).

Consider now a compact subset A of $\mathbb{R}^{n+1}$. For $x \in \mathbb{R}$, we denote by A_x the set $\{z \in \mathbb{R}^n \; ; \; (z, x) \in A\}$. Consider $x' \in \mathbb{R}$, $z \in \mathbb{R}^n$, and u such that $z \in A_{x'} + V_n(u)$. Then, by definition of $V_{n+1}(u)$, we have, for all $x \in \mathbb{R}$ that

$$(z, x) \in A + V_{n+1}\big(u + \xi(x - x')\big) .$$

Thus we have

$$h_A\big((z, x)\big) \leq \xi(x - x') + h_{A_{x'}}(z)$$

and hence

$$\exp h_A\big((z, x)\big) \leq \exp \xi(x - x') \exp h_{A_{x'}}(z) .$$

By the induction hypothesis, we have

$$\int_{\mathbb{R}^n} \exp h_A\big((z, x)\big) d\mu^n(z) \leq 2 \frac{e^{\xi(x - x')}}{\mu^n(A_{x'})} .$$

Thus

$$\int_{\mathbb{R}^n} \exp h_A\big((z, x)\big) d\mu^n(z) \leq 2 \inf_{x' \in \mathbb{R}} \frac{e^{\xi(x - x')}}{\mu^n(A_{x'})} .$$

By Fubini theorem, we have

$$\int_{\mathbf{R}^{n+1}} e^{h_A(y)} d\mu^{n+1}(y) \le 2 \int_{\mathbf{R}} \inf_{x' \in \mathbf{R}} \frac{e^{\xi(x-x')}}{\mu^n(A_{x'})} d\mu(x) . \tag{5.1}$$

Consider the function $g(x)$ given by $\exp g(x) = \mu^n(A_x)$. By Fubini theorem, we have $\int \exp g(x) d\mu(x) = \mu^{n+1}(A)$. On the other hand, the right hand side of (5.1) is

$$2 \int_{\mathbf{R}} e^{-\hat{g}(x)} d\mu(x)$$

where $\hat{g}(x) = \sup_{x' \in \mathbf{R}} (g(x') - \xi(x - x'))$. Thus to prove that $(H_n) \Rightarrow (H_{n+1})$, it suffices to show that

$$\int_{\mathbf{R}} e^{-\hat{g}(x)} d\mu(x) \int_{\mathbf{R}} e^{g(x)} d\mu(x) \le 1 .$$

The basic inequality (3.2) shows that this it true when g is non-decreasing. To reduce the general case to that case, consider the non-decreasing rearrangement f of g given by (2.4). Since $\mu(\{f \ge y\}) = \mu(\{g \ge y\})$ for all $y \in \mathbf{R}$, we have $\int_{\mathbf{R}} e^{g(x)} d\mu(x) = \int_{\mathbf{R}} e^{f(x)} d\mu(x)$. It suffices to show that for all t,

$$\mu(\{\hat{f}(x) > y\}) \le \mu(\{\hat{g}(x) > y\}) .$$

The argument to prove this is identical to that of Proposition 2.6. Theorem 1.2 is proved.

Finally, we prove that even a weak form of the concentration of measure property can hold only for powers of measures with exponential tails.

Proposition 5.1. *Consider a probability θ on $\mathbf{R}$, and its power θ^∞ on $\mathbf{R}^{\mathbf{N}}$. Assume that there exists $u_0 \in \mathbf{R}$ with the following property. For each Borel set $A \subset \mathbf{R}^{\mathbf{N}}$*

$$\theta^\infty(A) \ge 1/2 \Rightarrow \theta^\infty(A + u_0 B_\infty) \ge \frac{3}{4},$$

where $B_\infty = \{x \in \mathbf{R}^{\mathbf{N}}; \forall k \ge 1, |x_k| \le 1\}$. Then $\theta(\{|x| \ge u\}) \le K \exp(-u/K)$.

Comment. The choice of the number $\frac{3}{4}$ is not magical. Any number $> \frac{1}{2}$ would serve the same purpose.

Proof: Consider $u > 0$. Let n be the smallest integer such that $\theta([-u, u])^n \ge 1/2$. Set

$$A = \{x \in \mathbf{R}^{\mathbf{N}} ; \forall k \le n , |x_k| \le u\}$$

so that $\theta^\infty(A) \ge 1/2$. Then

$$A + u_0 B_\infty = \{x \in \mathbf{R}^{\mathbf{N}}; \forall k \le n, |x_k| \le u + u_0\} .$$

Thus

$$\theta^\infty(A + u_0 B_\infty) = \theta\big([-u - u_0, u + u_0]\big)^n \geq \tfrac{3}{4} \ .$$

Thus, we have

$$\theta\big(\{|x| > u\}\big) \leq 1 - \big(\tfrac{1}{2}\big)^{1/n} \Rightarrow \theta\big(\{|x| > u + u_0\}\big) \leq 1 - \big(\tfrac{3}{4}\big)^{1/n} \ .$$

Since $1 - \big(\tfrac{1}{2}\big)^{1/n}$ is of order $\tfrac{1}{n}\log 2$, while $1 - \big(\tfrac{3}{4}\big)^{1/n}$ is of order $\tfrac{1}{n}\log\tfrac{4}{3}$, it follows that for u large enough we have

$$\theta\big(\{|x| \geq u + u_0\}\big) \leq \gamma\theta\big(\{|x| \geq u\}\big) \ ,$$

where $\gamma < 1$. The result follows easily. $\qquad\qquad\square$

References

[B]   C. Borell, The Brunn-Minkovski inequality in Gauss space, Invent. Math. 30, 207-216 (1975).

[E1]   A. Ehrhard, Symétrisation dans l' espace de Gauss, Math. Scand. 53, 281-301 (1983).

[E2]   A. Ehrhard, Inégalités isopérimétriques et intégrales de Dirichlet Gaussiennes, Ann. Scient. Ec. Norm. Sup. 17, 317-332 (1984).

[H]   W. Hoeffding, Probability inequalities for sums of bounded random variables, J. Amer. Statist. Assoc. 58, 13-30 (1963).

[M]   V.D. Milman, The concentration phenomenon and linear structure of finite dimensional normed spaces, Proceedings of the International Congress of Mathematicians, Berkeley, 961-975 (1986).

[P]   G. Pisier, Probabilistic methods in the geometry of Banach spaces, Probability and Analysis, Varenna (Italy) 1985, Lecture Notes in Math. 1206, 167-241 (1986) Springer-Verlag.

[ST]   V.N. Sudakov, B.S. Tsirel'son, Extremal properties of half spaces for spherically invariant measures, J. Soviet Math. 9, 9-18 (1978), translated from Zap. Nauch. Sem. L.O.M.I. 41, 14-24 (1974).

[T]   M. Talagrand, Sur l'intégrabilité des vecteurs gaussiens, Z. Wahrscheinlichkeits theorie verw. Gebiete 68, 1-8 (1984).

PERMUTATIONS OF THE HAAR SYSTEM

Paul F.X. Müller*

Institut für Mathematik
J. Kepler Universität
Linz, Austria
and
Department of Theoretical Mathematics
The Weizmann Institute of Science
Rehovot, Israel

Abstract

Permutations which act on one level of the Haar system only, are considered. A short straightforward proof of a result due to E.M. Semyonov and B. Stöckert is given.

Let us briefly describe the setting in which we are working. $\mathcal{D}$ denotes the set of all dyadic intervals contained in the unit interval. $\pi : \mathcal{D} \to \mathcal{D}$ denotes a permutation of the dyadic intervals. The operator induced by π is determined by the equation

$$T_\pi h_I = h_{\pi(I)}$$

where h_I denotes the L_∞-normalised Haar function supported on the dyadic interval I. To study these operators on L_p E.M. Semyonov introduced the parameter,

$$K = \sup \left\{ \frac{|\pi^{-1}(\mathcal{B})^*|}{|\mathcal{B}^*|} : \mathcal{B} \subseteq \mathcal{D} \right\}$$

where for example $\mathcal{B}^*$ denotes the pointset covered by the collection $\mathcal{B}$.

E.M. Semyonov and B. Stöckert proved the following result.

Theorem 1. *If for every $i \in \mathcal{D}$ we have $|\pi(I)| = |I|$ and $K < \infty$ then for $2 \le p < \infty$ the operator T_π is bounded on L_p.*

We will obtain this result from

---

* Supported by E. Schrödinger auslandsstipendium PR.Nr J0458-PHY

Theorem 2. *If for every $I \in \mathcal{D}$ we have $|\pi(I)| = |I|$ and $K < \infty$ then the operator T_π is bounded on dyadic-BMO.*

Proof: Recall that for formal series $f = \sum_{I \in \mathcal{D}} a_I h_I$ the dyadic - BMO norm is given by

$$\left(\sup_{\mathcal{B}} \frac{1}{|\mathcal{B}^*|} \sum_{I \in \mathcal{B}} a_I^2 |I| \right)^{\frac{1}{2}}$$

where the supremum is extended over all collections of dyadic intervals $\mathcal{B}$. We fix now $x = \sum x_I h_I$ and obtain $T_\pi x = \sum x_{\pi^{-1}(I)} h_I$. Then choose $\mathcal{B} \subseteq \mathcal{D}$ such that

$$\frac{1}{2} \|T_\pi x\|_{BMO}^2 \leq \frac{1}{|\mathcal{B}^*|} \sum_{\mathcal{B}} x_{\pi^{-1}(I)}^2 |I|$$

By hypothesis this expression equals with

$$\frac{1}{|\mathcal{B}^*|} \sum_{\mathcal{B}} x_{\pi^{-1}(I)}^2 |\pi^{-1}(I)|$$

Which equals trivially with

$$\frac{|\pi^{-1}(\mathcal{B})^*|}{|\mathcal{B}^*|} \frac{1}{|\pi^{-1}(\mathcal{B})^*|} \sum_{\mathcal{B}} x_{\pi^{-1}(I)}^2 |\pi^{-1}(I)|$$

The last expression is of course bounded by $K \|x\|_{BMO}^2$. This finishes the proof of Theorem 2.

Remark. 1) For every permutation which satisfies $|\pi(I)| = |(I)|$ there exist $\mathcal{E} \subseteq \mathcal{D}$ and $x \in BMO$ for which the above chain of inequalities can be reversed. Hence for such permutations the condition $K < \infty$ *is implied by* the boundedness of T_π.

2) As T_π is bounded on L_2 we obtain from [2] Corollary 2 p. 60 that T_π is bounded on L_p for $2 < p < \infty$. We thus obtained Theorem 1 from Theorem 2 by interpolation.

References

[1]   R.R. Coifman and G. Weiss. Extensions of Hardy spaces and their use in Analysis, Bull. Amer. Math. Soc. 83 (1977).

[2]   S. Janson and P.W. Jones Interpolation between H^p spaces: The complex method, Journal o. Functional Analysis 48 (1982).

[3]   E.M. Semyonov and B. Stöckert. Haar system rearrangement in the spaces L_p, Analysis Mathematika 7 (1981).

ON THE DISTRIBUTION OF POLYNOMIALS ON HIGH DIMENSIONAL CONVEX SETS

J. Bourgain

IHES, France
and
University of Illinois, USA

1. Introduction

In this paper we will prove results on convex sets K in R^n where the particular emphasis will lie on the role of the dimension n. The statements are either independent of K and dimension free or one will seek for an estimate depending on n which is as good as possible. If one fixes the dimension n, the results are essentially trivial.

Our main result deals with the behaviour of polynomials on general convex sets.

Theorem 1.1. *For any positive integer d and number $p < \infty$, there is a constant $c(d, p)$ such that if f is any polynomial in n variables, i.e. $f = f(x_1, \ldots, x_n)$ of degree d and K any convex body in R^n of volume $\operatorname{Vol} K = 1$, then*

$$\left(\int_K |f(x)|^p dx \right)^{1/p} \leq c(d, p) \int_K |f(x)| dx . \tag{1.2}$$

In fact, a more precise result holds.

Theorem 1.3. *Under the hypothesis of Theorem 1.1, there is an inequality*

$$\|f\|_{L^\psi(K, dx)} \leq C \|f\|_{L^1(K, dx)} , \tag{1.4}$$

where L^ψ refers to the Orlicz space with Orlicz function

$$\psi(t) = \exp(t^{c/d}) - 1 . \tag{1.5}$$

The constants C, c appearing in (4),(5) are absolute.

Of course, Theorem 1.3 implies Theorem 1.1. These facts may be seen as a rather far reaching generalization of the classical probabilistic results on moment equivalence for Walsh functions on the Cantor group $\{1, -1\}^{\mathsf{Z}_+}$ of a given order d. In this case, (1.5) holds with $c = 2$.

The same statement also remains valid if one specifies K to be the n-dimensional euclidean ball or the n-dimensional cube.

Theorem 1.1 answers affirmatively a problem and conjecture expressed by V. Milman to the author[*]. The particular case $d = 1$, thus f linear, was already known (proved by Gromov and Milman) and plays a role in certain problems (see [Bo] and also the appendix at the end of this paper).

In the case $d = 1$, (1.5) holds with $c = 1$. This value for c may not be improved in general, as the example of the simplex $\left[\sum_1^n |x_i| \leq 1 \right]$ shows.

The appendix of the paper deals with a lower bound for the codimension-1 central sections of a convex 0-symmetric body K in $\mathbf{R}^n$. The result obtained is probably not final but has been mentioned in the literature (see [M-P], [B-M- M-P]) and is the best known general estimate to date.

Theorem 1.6. *Let K be a convex centrally symmetric body in $\mathbf{R}^n$, $\mathrm{Vol}\, K = 1$. There is a hyperplane H such that*

$$\mathrm{Vol}_{n-1}(K \cap H) \gtrsim n^{-1/4} . \tag{1.7}$$

(We use the notation $\gtrsim$ to indicate a possible factor which only grows logarithmically in dimension n.)

A previously known estimate, based on the euclidean distance information, stated that

$$\mathrm{Vol}_{n-1}(K \cap H) > cn^{-1/2} \tag{1.8}$$

(see [M-P]). It is a conjecture that

$$\sup_{H \text{ hyperplane}} \mathrm{Vol}_{n-1}(K \cap H) > c , \tag{1.9}$$

for some universal constant c. This fact was indeed verified for many classes of bodies.

Observe that conversely

$$\min_{H \text{ hyperplane}} \mathrm{Vol}_{n-1}(K \cap H) < c , \tag{1.10}$$

for some absolute constant C.

There is an amazing variety of reformulations of conjecture (1.9). The reader should consult [M-P] for a complete discussion.

---

[*] Tel Aviv, June 1989

The proof of (1.7) depends ultimately on Theorem 3 applied in the linear case, i.e. with Orlicz function

$$\psi(t) = (\exp t) - 1 . \tag{1.11}$$

Besides, one uses certain ingredients from finite dimensional Banach space theory, namely the existence of certain ellipsoids. It would be difficult to be selfcontained in our exposition here and the reader is referred to [M-S] for matters related to the "asymptotic theory of normed spaces" or G. Pisier's book [P1].

Summary.

1. Introduction
2. The Knothe parametrization
3. Distribution of polynomials in 1 variable
4. Proof of Theorem 1.3
5. Appendix: On the section problem for convex bodies.

2. The Knothe Parametrization

In this section, we recall the statement and construction of the Knothe mapping [Kn].

Proposition 2.1. *Fix a coordinate system* $x_1, \ldots, x_n$ *in* $\mathbf{R}^n$, *let* A, B *be open subsets of* $\mathbf{R}^n$ *and assume* B *convex. There exists a one-to-one map* φ *from* A *onto* B *with following properties*

(2.2) *The map* φ *is triangular, i.e.* $\varphi_i = \varphi_i(x_1, \ldots, x_i)$

(2.3) *The partial derivatives* $\frac{\partial \varphi_i}{\partial x_i}$ *are non-negative on* A *and the Jacobian*

$$J(x) = \mathrm{Det}(D\varphi) = \prod_{i=1}^{n} \frac{\partial \varphi_i}{\partial x_i} \tag{2.4}$$

satisfies

$$J(x) = \frac{\mathrm{Vol}\, B}{\mathrm{Vol}\, A} \quad \text{for} \quad x \in A . \tag{2.5}$$

Remarks.

(2.6) As the proof will show, an omission of the convexity hypothesis on B may lead to singularity sets for φ. The convexity of B will ensure in particular that any set of the form $B \cap [x_i = a_i \mid i \in I]$, for some $I \subset \{1, \ldots, n\}$, is connected.

(2.7) The Knothe mapping is a powerful tool in investigating almost-equality situations in Brunn's theorem $\mathrm{Vol}(A + B)^{1/n} \geq (\mathrm{Vol}\, A)^{1/n} + (\mathrm{Vol}\, B)^{1/n}$, see for instance [B-L].

Proof of (2.1).

Define for $i = 1, \ldots, n$ and $s = (s_1, \ldots, s_i) \in \mathbf{R}^i$ the section

$$A_s = \{ y \in \mathbf{R}^{n-i} \mid (s, y) \in A \} , \tag{2.8}$$

and similarly for B.

The function $\varphi_1(_1)$ is defined by the relation

$$\frac{\mathrm{Vol}_n B}{\mathrm{Vol}_n A} \cdot \int_{-\infty}^{x_1} \mathrm{Vol}_{n-1}(A_{s_1}) ds_1 = \int_{-\infty}^{\varphi_1(x_1)} \mathrm{Vol}_{n-1}(B_{t_1}) dt_1 , \tag{2.9}$$

for x_1 such that $A_{x_1} \neq \emptyset$. Hence

$$\frac{\partial \varphi_1}{\partial x_1} = \frac{\mathrm{Vol}_n B}{\mathrm{Vol}_n A} \cdot \frac{\mathrm{Vol}_{n-1}(A_{x_1})}{\mathrm{Vol}_{n-1}(B_{\varphi_1(x_1)})} , \tag{2.10}$$

where clearly $B_{\varphi_1(x_1)} \neq \emptyset$.

Proceeding by induction, one puts in general

$$\frac{\mathrm{Vol}_{n-i}(B_{\varphi_1(x_1), \ldots, \varphi_i(x_1, \ldots, x_i)})}{\mathrm{Vol}_{n-i}(A_{x_1, \ldots, x_i})} \int_{-\infty}^{x_{i+1}} \mathrm{Vol}_{n-i-1}(A_{x_1, \ldots, x_i, s_{i+1}}) ds_{i+1} =$$

$$\int_{-\infty}^{\varphi_{i+1}(x_1, \ldots, x_{i+1})} \mathrm{Vol}_{n-i-1}(B_{\varphi_1(x_1), \ldots, \varphi_i(x_1, \ldots, x_i), t_{i+1}}) dt_{i+1} , \tag{2.11}$$

for $i < n$ and $x_1, \ldots, x_i$ satisfying $A_{x_1, \ldots, x_{i+1}} \neq \emptyset$. Again

$$\frac{\partial \varphi_{i+1}}{\partial x_{i+1}} = \frac{\mathrm{Vol}_{n-i}(B_{\varphi_1(x), \ldots, \varphi_i(x)})}{\mathrm{Vol}_{n-i}(A_{x_1, \ldots, x_i})} \cdot \frac{\mathrm{Vol}_{n-i-1}(A_{x_1, \ldots, x_{i+1}})}{\mathrm{Vol}_{n-i-1}(B_{\varphi_1(x), \ldots, \varphi_{i+1}(x)})} \tag{2.12}$$

where for $i = n - 1$ the second factor in (2.12) is dropped. Clearly (2.4) holds and (2.5) is implied by (2.10) and equalifies (2.12).

3. Distribution of Polynomials in One Variable

The proofs of Theorems 1.1 and 1.3 eventually reduce to the general behaviour of a 1-variable polynomial $p(t)$ of degree d on the interval $[0, 1]$. The following lemma will be used.

Lemma 3.1. *There is a constant $c_1 > 1$ such that if $p(t) = \sum_0^d a_j t^j$ is an arbitrary polynomial and $\|p\|_\infty = \sup\limits_{t \in [0,1]} |p(t)|$, then*

$$\mathrm{mes}\, [t \in [0, 1] \mid |p(t)| \geq c_1^{-d} \|p\|_\infty] > \tfrac{1}{2} . \tag{3.2}$$

Clearly, Lemma 3.1 is a consequence of the following.

Lemma 3.3. *There is a constant c such that for $p(t)$ as above*

$$\int_0^1 |p(t)|^{-\frac{1}{2d}} dt \leq c \|p\|_\infty^{-\frac{1}{2d}} , \tag{3.4}$$

by a simple use of Tchébycheff's inequality.

Proof of Lemma 3.3.   Let $z_1, \ldots, z_d$ be the zero's of p (in the complex plane) and write

$$p(t) = A \prod_1^d (t - z_j) . \tag{3.4}$$

Since

$$\int_0^1 |t - z|^{-1/2} dt < c(1 + |z|)^{-1/2} \tag{3.5}$$

for any $z \in \mathbb{C}$, it follows from Hölders inequality that

$$\int_0^1 |p(t)|^{-\frac{1}{2d}} dt = A^{-\frac{1}{2d}} \int_0^1 \prod_1^d |t - z_j|^{-\frac{1}{2d}} dt$$

$$\leq A^{-\frac{1}{2d}} \left\{ \prod_1^d \int_0^1 |t - z_j|^{-1/2} dt \right\}^{1/d} \tag{3.6}$$

$$\leq c \left[A \prod_1^d (1 + |z_j|) \right]^{-\frac{1}{2d}} \tag{3.7}$$

$$\leq c \|p\|_\infty^{-\frac{1}{2d}} . \tag{3.8}$$

Inequality (3.8) is indeed immediate from (3.4).

4. Proof of Theorem 1.3

Let $f = f(x_1, \ldots, x_n)$ and K be as in Theorem 1.1. Let

$$\int_K |f(x)| dx = 1 . \tag{4.1}$$

To bound $\|f\|_{L^\psi(K)}$, one needs to prove the corresponding distributional inequality. Put

$$K_\lambda = [x \in K \mid |f(x)| > \lambda] , \tag{4.2}$$

assuming $\lambda < \|f\|_\infty \equiv \sup_{x \in K} |f(x)|$.

Application of Proposition 2.1 with $A = K_\lambda$, $B = K$ yields a map $\varphi : K_\lambda \to K$ with the properties (2.2), (2.3). Fixing $x \in K_\lambda$, one considers the polynomial in t

$$p_x(t) = f(tx + (1 - t)\varphi(x)) \tag{4.3}$$

which is of degree at most d. Since by construction

$$\sup_{t \in [0,1]} |p_x(t)| \geq |p_x(1)| = |f(x)|\lambda , \tag{4.4}$$

(3.2) yields

$$\text{mes}\left[t \in [0,1] \mid |p_x(t)| > c_1^{-d}\lambda\right] > \tfrac{1}{2} \tag{4.5}$$

Hence, integrating on K_λ,

$$\frac{1}{\text{mes}\,K_\lambda} \int_{K_\lambda} \text{mes}\left[t \in [0,1] \mid |p_x(t)| > c_1^{-d}\lambda\right] dx > \tfrac{1}{2}\;. \tag{4.6}$$

It easily follows then from Fubini's theorem that for $0 \le t \le \tfrac{2}{3}$ one has

$$\text{mes}\left[x \in K_\lambda \mid |p_x(t)| > c_1^{-d}\lambda\right] > \tfrac{1}{4}\,\text{mes}\,K_\lambda\;. \tag{4.7}$$

Clearly, because of (2.3), the map $tx + (1-t)\varphi(x)$ is one-to-one on K_λ and maps the set

$$S = \left\{x \in K_\lambda \mid |p_x(t)| > c_1^{-d}\lambda\right\}, \tag{4.8}$$

on a set S' which measures

$$\text{mes}\,S' = \int_S \text{Det}\,(tId + (1-t)D\varphi) = \int_S \prod_{i=1}^{n}\left(t + (1-t)\frac{d\varphi_i}{\partial x_i}\right)$$

$$\ge \int_S \left(\prod_1^n \frac{\partial\varphi_i}{\partial x_i}\right)^{1-t}$$

$$= \int_S J(x)^{1-t}$$

$$= \left(\frac{\text{mes}\,K}{\text{mes}\,K_\lambda}\right)^{1-t} \text{Vol}\,S$$

$$\ge \tfrac{1}{4}(\text{mes}\,K_\lambda)^t\;, \tag{4.9}$$

by (2.4), (2.5), (4.7).

Again by construction, $S' \subset K_{c_1^{-d}\lambda}$. Since $t \le \tfrac{2}{3}$, (4.9) therefore implies

$$\text{mes}\,K_{c_1^{-d}\lambda} \ge \tfrac{1}{4}(\text{mes}\,K_\lambda)^{2/3}\;. \tag{4.10}$$

It remains to iterate (4.10). Letting λ be sufficiently large, choose r such that

$$c_1^{rd} \le \frac{\lambda}{1000} < c_1^{(r+1)d}\;. \tag{4.11}$$

Applying r times (4.10) one gets

$$\text{mes}\,K_{c_1^{-rd}\lambda} \ge \frac{1}{4^3}(\text{mes}\,K_\lambda)^{\left(\frac{2}{3}\right)^r}\;, \tag{4.12}$$

where, by (4.11) and normalization (4.1)

$$\text{mes } K_{c_1^{-rd}\lambda} \leq \text{mes } K_{1000} < \frac{1}{1000} \, . \tag{4.13}$$

Hence, by (4.12), (4.13)

$$\text{mes } K_\lambda \leq \left(\frac{64}{1000}\right)^{\left(\frac{3}{2}\right)^r} < \exp(-\lambda^{cd}) \, , \tag{4.14}$$

for some absolute constant $c > 0$.

Deriving Theorem 1.3 from this distribution inequality for f is standard.

5. Appendix: The Section Problem for Convex Bodies

In this section we give a proof of Theorem 1.6. The argument is not self contained and the reader may wish to consult [M-P], [P1] for further details on some concepts mentioned below.

Let K be a convex symmetric body in $\mathbb{R}^n$. The problem being affinely invariant, one may assume K appears in its isotropic position, i.e.

$$\int_K |\langle x, \theta \rangle|^2 dx = L^2 \quad \text{for all} \ \ \theta \in S^{n-1} \, , \tag{5.1}$$

where S^{n-1} denotes the unit sphere. In this position, all codimension-1 central sections of K have approximately Volume $\frac{1}{L}$, in the sense that

$$\frac{c_2}{L} \leq \text{Vol}_{n-1}(K \cap H) \leq \frac{c_1}{L} \, , \tag{5.2}$$

where c_1, c_2 are absolute.

It is known that $L \geq c$ for some absolute constant and our purpose is thus to obtain an upper estimate

$$L \lesssim n^{1/4} \, , \tag{5.3}$$

where $\lesssim$ denotes inequality up to a factor logarithmic in the dimension n.

Besides the Binet Ellipsoid involved in defining the isotropic position, essential use will also be made of the so-called ℓ- ellipsoid of K, which is a concept of finite dimensional Banach space theory. This amounts to the existence of positive scalars $\lambda_1, \ldots, \lambda_n$ such that

$$\int \left\| \sum_1^{\tilde{}} \lambda_i g_i(w) e_i \right\|_* dw \lesssim n \tag{5.4}$$

$$\int \left\| \sum_1^{\tilde{}} \lambda_i^{-1} g_i(w) e_i \right\| dw \lesssim 1 \, . \tag{5.5}$$

Here $(e_i)_{1 \leq i \leq n}$ are the unit vectors of $\mathbb{R}^n$, $\{g_i\}_{i=1}^n$ a sequence of independent L^2-normalized Gaussians and $\| \ \|$ (resp. $\| \ \|_*$) stand for the norm induced by K (resp. K°) on $\mathbb{R}^n$.

One clearly has by (5.5) and volume computation

$$\left\{ \frac{\operatorname{Vol} K}{\left(\prod_1^n \lambda_i^{-1} \right) \operatorname{Vol} B_n} \right\}^{1/n} \sim \left\{ \int \left(\frac{1}{\sqrt{n}} \| \sum \lambda_i^{-1} g_i(w) e_i \| \right)^{-n} dw \right\}^{1/n} \tag{5.6}$$

$$\geq \frac{\sqrt{n}}{\int \| \sum \lambda_i^{-1} g_i(w) e_i \| dw} \gtrsim \sqrt{n} \ , \tag{5.7}$$

where B_n denotes the unit ball in $\mathbb{R}^n$. Hence

$$\left(\prod_1^n \lambda_i^{-1} \right)^{1/n} \lesssim 1 \tag{5.8}$$

and therefore

$$\frac{1}{n} \sum_1^n \lambda_i \geq \left(\prod_1^n \lambda_i \right)^{1/n} \gtrsim 1 \ . \tag{5.9}$$

Consider next the integral

$$\int_K \left\| \sum_1^n \lambda_i \frac{x_i}{L} e_i \right\|_* dx \tag{5.10}$$

replacing in (5.4) the Gaussians g_i by the function $\frac{\langle x, e_i \rangle}{L}$ on K. Obviously

$$(5.10) \geq \int_K \left(\sum_1^N \lambda_i \frac{x_i^2}{L} \right) dx \gtrsim Ln \ , \tag{5.11}$$

invoking (5.1),(5.9).

Our final purpose is to relate (5.10) and the Gaussian expression in order to get an upper bound from (5.4). This is the weak point in the argument. One has the general information

$$\left\| \frac{\langle x, \theta \rangle}{L |\theta|} \right\|_{L^{\psi_1}(K)} \leq C \tag{5.12}$$

where $\psi_1(x) = e^x - 1$. Denoting $\psi_2(x) = e^{x^2} - 1$ the Gaussian Orlicz function, one therefore has for $K' \subset K$, $\operatorname{Vol} K' \sim 1$

$$\left\| \frac{\langle x, \theta \rangle}{L} \right\|_{L^{\psi_2}(K')} \leq c \left(\frac{\operatorname{diam} K'}{L} \right)^{1/2} \ , \tag{5.13}$$

for $\theta \in S^{n-1}$.

Observe that

$$\int_K \left(\sum_1^n x_i^2 \right) dx = nL^2 . \tag{5.14}$$

Defining for $t > 1$

$$K_t = K \cap \{ x \in \mathbb{R}^n \mid |x| < t\sqrt{n}\, L \} , \tag{5.15}$$

it follows from Theorem $1.3^{(*)}$ that

$$\mathrm{mes}(K' \backslash K_t) < e^{-t^c} , \tag{5.16}$$

uniformly for $t \to \infty$. Hence, choosing t sufficiently large (corresponding again to an absolute constant) one may ensure that

$$\int_{K'} x_i^2 dx > \tfrac{1}{2} L^2 \quad \text{for} \quad 1 \le i \le n , \tag{5.17}$$

letting $K' = K_t$. Thus (5.1) remains valid for K replaced by K', while (5.13) becomes

$$\left\| \left\langle \frac{x}{L}, \theta \right\rangle \right\|_{L^{\psi_2}(K)} \le C n^{1/4} . \tag{5.18}$$

We will use the following fact.

Lemma 5.19. *Let $\{h_i\}_{i=1}^n$ be functions on a probability space satisfying for $\theta \in S^{n-1}$*

$$\left\| \sum_1^n \theta_i h_i \right\|_{L^{\psi_2}(dx)} \le M . \tag{5.20}$$

Let $\mathcal{E}$ be a bounded subset of $\mathbb{R}^n$. Then

$$\int \sup_{\theta \in \mathcal{E}} \left| \sum_1^n h_i(x)\theta_i \right| dx \le M_1 \int \sup_{\theta \in \mathcal{E}} \left| \sum_1^n g_i(w)\theta_i \right| dw , \tag{5.21}$$

where $M_1 \sim M \cdot \log(n + \mathrm{diam}\, \mathcal{E})$.

(5.21) is obtained by straightforward entropy considerations similar to those in the Dudley-Fernique type inequalities and using Sudakov's minoration for the Gaussian average (cf. [Fe]). We leave the details as exercise to the reader.

$^{(*)}$ In this case the result was already known, since we are dealing with a convex function.

Remark 5.22. G. Pisier [P2] observed to me that in fact (5.21) holds with $M_1 \sim M$. This fact is deeper and is based on M. Talagrand's majorizing measure theorem [T] and work of C. Preston [Pr]. This improvement is of course irrelevant here.

We can now complete the proof of Theorem 1.6. Considering the functions $h_i(x) = \frac{x_i}{L}$ on K', (5.18) yields (5.20) with $M \sim n^{1/4}$. Taking for $\mathcal{E}$ the image of K under the diagonal map $e_i \mapsto \lambda_i e_i$, it follows from (5.21), (5.4) that

$$\int_{K'} \left\| \sum \lambda_i \frac{x_i}{L} e_i \right\|_* dx \lesssim n^{5/4} . \tag{5.23}$$

As observed above, also

$$\int_{K'} \left\| \sum \lambda_i \frac{x_i}{L} e_i \right\|_* dx \gtrsim Ln , \tag{5.24}$$

implying the upper bound on L.

Remarks.

(5.25) The previous argument yields in fact $L < cn^{1/4} \log n$, taking in particular (5.22) into consideration

(5.26) It also follows that L_K is bounded if affine functions have equivalent $L^1(K)$ and $L^{\psi_2}(K)$-norms (this is the case for the ball and the cube for instance). Boundedness of L_K is, however, known for other classes of convex symmetric bodies, for instance the zonoids and their duals.

References

[Bo]   J. Bourgain, On high dimensional maximal functions associated to convex bodies, Amer. J. Math 108 (1986), 1467-1476.

[B-L]   J. Bourgain, J. Lindenstrauss, Projection bodies, Springer LNM 1317.

[B-M-M-P] J. Bourgain, M. Meyer, V. Milman, A. Pajor, On a geometric inequality, Springer LNM 1317, 271-282.

[Fe]   X. Fernique, Régularité des trajectoires des fonctions aléatoires Gaussiennes, Springer LNM 480 (1975), 1-96.

[Gr-M]   M. Gromov, V. Milman, Brunn theorem and a concentration of volume of convex bodies, GAFA Seminar Notes, Tel Aviv University, Israel 1983-84.

[Kn]   H. Knothe, Contributions to the theory of convex bodies, Michigan Math. J. 4 (1957), 39-52.

[M-P]   V. Milman, A. Pajor, Isotropic position and inertia ellipsoids and zonoids of the unit ball of a normed n-dimensional space, Springer LNM 1376, 64-104.

137

[M-S]   V. Milman, G. Schechtman, Asymptotic theory of finite dimensional normed spaces, Springer LNM 1200.

[P1]   G. Pisier, The Volume of Convex Bodies and Banach Space Geometry, Cambridge UP, 1989.

[P2]   G. Pisier, Private communication.

[Pr]   C. Preston, Banach spaces arising from some integral inequalities', Indiana U. Math. J. 20 (1971), 997-1015.

[T]   M. Talagrand, Regularity of Gaussian processes, Acta Math. 159 (1987), 99-149.

ON COVERING A SET IN R^N BY
BALLS OF THE SAME DIAMETER

J. Bourgain
I.H.E.S.
Bures sur Yvette
France

J. Lindenstrauss
Hebrew University
Jerusalem
Israel

In this note we report on some progress concerning the question of how many balls of radius $\frac{1}{2}$ are needed in order to cover a set of diameter 1 in the n-dimensional Euclidean space R^n. Our interest is in asymptotic results as $n \to \infty$.

As far as we know the previously known facts concerning this problem are the following: Every set of diameter 1 in R^n can be covered, for every $\varepsilon > 0$ and $n > n(\varepsilon)$, by $(\sqrt{2}+\varepsilon)^n$ balls of radius $\frac{1}{2}$. This fact is an immediate consequence of the classical fact that a set of diameter 1 is contained in a ball of radius $1/\sqrt{2}$ and of the result of Rogers [R] that a ball of radius r in R^n can be covered by $(r/\rho+\varepsilon)^n$ balls of radius $\rho < r$ if $n > n(\varepsilon)$. (This consequence of Rogers' result was apparently first pointed out explicitly by Schramm [S1].) On the other hand it is noted in [D] that an exponential number of balls is needed. More precisely, it is stated in [D] that there is a subset of R^n of diameter 1 which cannot be covered by less than $(1,003)^n$ balls of radius $\frac{1}{2}$.

In the present note, we improve the lower and upper bounds.

Proposition. *Let c_n denote the smallest integer such that any set in R^n of diameter 1 can be covered by c_n balls of the same diameter. Then for every $\varepsilon > 0$ and $n > n(\varepsilon)$*

$$\left(\sqrt{9/8} - \varepsilon\right)^n \le c_n \le \left(\sqrt{3/2} + \varepsilon\right)^n . \tag{1}$$

Both bounds in (1) are apparently not sharp. In fact, for the lower bound it will be proved below (see (8)) that $\sqrt{9/8} \approx 1,0607$ can be replaced, e.g. by $1,0645$. The proof of the lower bound in (1) is extremely easy; the main point of this note is the proof of the upper bound in (1). As will be evident from the proof, the crucial test case for checking whether the upper

---

Supported in part by a B.S.F. grant

bound in (1) can be improved is that of a set of diameter 1 contained in the sphere of radius $\sqrt{3/8}$ (i.e. $\{x \; ; \; \|x\| = \sqrt{3/8}\}$). It seems that further improvement of the bounds for c_n depends on questions related to the problem of dense packings of caps on spheres (in the sense of [KL]).

Our study of the question considered in this note was partially motivated by the well known Borsuk conjecture (see, e.g. [G]). It follows directly from the proposition that any set of diameter 1 in R^n can be covered by $\left(\sqrt{3/2} + \varepsilon\right)^n$ sets of diameter $\leq \left(1 - \varepsilon/2\right)$ for $n \geq n(\varepsilon)$. That a set of diameter 1 in R^n can be covered by $\left(\sqrt{3/2} + \varepsilon\right)^n$ sets of diameter < 1 is already known. This was proved by Schramm [S2] who showed that any set A of constant width in R^n can be covered by $\left(\sqrt{3/2} + \varepsilon\right)^n$ sets of the form $\alpha_i + \lambda_i A$ with $\lambda_i < 1$ for all i.

Notations. All the distances mentioned below are taken with respect to the Euclidean norm $\| \cdot \|$ in R^n. The set $\{x \; ; \; x \in R^n \, , \; \|x\| = r\}$ is denoted by S_r^{n-1}. If $0 < \|x_0\| < r$ the set

$$\|x \in S_r^{n-1} \; ; \; \|x - x_0\|^2 \leq r^2 - \|x_0\|^2\}$$

is a cap of S_r^{n-1} of diameter $2(r^2 - \|x_0\|^2)^{1/2}$. The point $rx_0/\|x_0\|$ of S_r^{n-1} is called the center of the cap. The point x_0 itself will be called the "metric center" of the cap.

The cardinality of a finite set B will be denoted by $|B|$.

For the proof of the proposition we need two lemmas.

Lemma 1. *Let $5/9 \leq r \leq 1$ and let $\varepsilon > 0$. Then for suitable $\delta = \delta(\varepsilon) > 0$ and for $n > n(\varepsilon)$ there exist covers*

$$\mathcal{A}_k = \{C_{k,m}\}_{m=1}^{M_k} \, , \qquad k = 0, 1, 2, \ldots, k_0 \tag{2}$$

of S_r^{n-1} by spherical caps $C_{k,m}$ of diameter $2\rho_k = (1 + \delta)^k$ with

$$M_k \leq \left((r + \varepsilon)/\rho_k\right)^n \, , \qquad 0 \leq k \leq k_0 \; ; \; (1 + \delta)^{k_0} \geq 9r/5 \tag{3}$$

and so that for each $k \geq 1$ and m

$$C_{k,m} \subset \bigcup_{j \in B_{k,m}} C_{k-1,j} \, , \qquad |B_{k,m}| \leq (1 + \varepsilon)^n \, , \tag{4}$$

Proof: We show first that any cap C of S_r^{n-1} of diameter 2ρ can be covered by $(1 + \delta + 25\delta^2)^n$ caps of diameter $2\rho/(1 + \delta)$ if $\frac{1}{2} \leq \rho \leq 9r/10$, $\delta > 0$ is small enough and $n > n(\delta)$. To do this we first pick C_0 to be the cap concentric with C and of diameter $2\rho/(1 + \delta)$. The boundary ∂C_0 (which up to a rigid motion is $S_{\rho/(1+\delta)}^{n-2}$) can be covered, in view of the result of Rogers

[R] mentioned above, by at most $\ell = (1 + \delta + 25\delta^2)^n - 1$ $(n-2)$-dimensional caps $\{C_j'\}_{j=1}^{\ell}$ of diameter $2\rho(1 - 2\delta - 20\delta^2)$. We used here the fact that for small enough δ

$$1 + \delta + 25\delta^2 > \left[(1+\delta)(1 - 2\delta - 20\delta^2)\right]^{-1} .$$

Any point of $C \setminus C_0$ differs from a suitable point in ∂C_0 by a vector whose component in the hyperplane determined by ∂C_0 is of length $\leq \rho - \rho/(1+\delta) = \delta\rho/(1+\delta)$, and whose component in the orthogonal direction is at most

$$\left(r^2 - \rho^2/(1+\delta)^2\right)^{1/2} - (r^2 - \rho^2)^{1/2} \leq 6\delta\rho ,$$

for small enough δ (here we used that $\frac{1}{2} \leq \rho \leq 9r/10$).

Since

$$\left(1 - 2\delta - 20\delta^2 + \delta/(1+\delta)\right)^2 + 36\delta^2 \leq 1/(1+\delta)^2 ,$$

we deduce that if $\{C_j\}_{j=1}^{\ell}$ are the caps of S_r^{n-1} with diameter $2\rho/(1+\delta)$ and the same metric centers as $\{C_j'\}_{j=1}^{\ell}$ then $\{C_j\}_{j=0}^{\ell}$ cover C as required.

Having verified the assertion made above we pass to the inductive construction of the coverings $\mathcal{A}_k$, starting with the covering with the largest caps (corresponding to $k = k_0$). We assume as we clearly may that δ and k_0 are such that $(1+\delta)^{k_0} = 9r/5$. The cover $\mathcal{A}_{k_0}$ is chosen so that M_{k_0} satisfies (3) for $k = k_0$; this can be done in view of Rogers' result. Having constructed $\mathcal{A}_k$ for some $k \geq 1$ we use the assertion proved above to construct for every m caps $\{C_{k-1,j}\}_{j \in B_{k,m}}$ of diameter $2\rho_k/(1+\delta) = 2\rho_{k-1}$ which cover $C_{k,m}$ and with

$$|B_{k,m}| \leq (1 + \delta + 25\delta^2)^n \leq (1+\varepsilon)^n$$

if $\delta = \delta(\varepsilon)$ is small enough, and put $\mathcal{A}_{k-1} = \{C_{k-1,j}\}_{j \in \bigcup_m B_{k,m}}$. We clearly have that (4) holds. For M_k we get by the construction the estimate

$$M_k \leq M_{k_0}(1 + \delta + 25\delta^2)^{(k_0 - k)n} .$$

Since $(1+\delta)^{k_0} \leq 2$ we deduce that for given $\varepsilon > 0$ there is a choice of $\delta(\varepsilon) > 0$ so that (3) holds for all $0 \leq k \leq k_0$. $\quad\square$

Lemma 2. Let K be a subset of diameter 1 of S_r^{n-1} with $r > \sqrt{3/8}$. Let C_1 and C_2 be two caps of diameter $2\rho < 2r$ on S_r^{n-1}. Assume that $C_1 \cap K$ and $C_2 \cap K$ are both not contained in the union of two caps of S_r^{n-1} of diameter 1. Then the distance between the centers of C_1 and C_2 is at most $2d(r,\rho)$ where

$$d(r,\rho) = \frac{r\left(2^{-3/2}(r^2 - \rho^2)^{1/2} + (r^2 - 3/8)^{1/2}(\rho^2 - 1/4)^{1/2}\right)}{r^2 - 1/4} \tag{5}$$

Proof: We choose the coordinate system so that the center of C_1 has coordinates $\left(\sqrt{r^2 - d^2}, d, 0, \dots\right)$ and that of C_2 has coordinates $\left(\sqrt{r^2 - d^2}, -d, 0, \dots\right)$. We remove from C_1 the "bottom part" for diameter 1, i.e. the ball $B\left(q_1, \frac{1}{2}\right)$ with center $q_1 = \left(\sqrt{r^2 - \frac{1}{4} - \alpha^2}, \alpha, 0, \dots\right)$ are radius $\frac{1}{2}$ (see the picture)

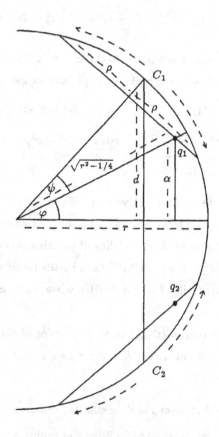

It is easy to check that any point in $C_1 \backslash B\left(q_1, \frac{1}{2}\right)$ has a second coordinate $\geq \alpha$. Since $C_1 \cap K$ is not contained in two balls of radius $\frac{1}{2}$ the closure of $K \cap C_1 \backslash B\left(q_1, \frac{1}{2}\right)$ contains a set $\{x_j\}_{j=1}^m$ so that the ball of minimal radius containing these points has radius $t_1 > \frac{1}{2}$ and so that all the x_j are on the boundary of this ball. The center of this ball, say u, can be represented as

$u = \sum_{j=1}^{m} \lambda_j x_j$ with $\lambda_j \geq 0$ and $\sum_j \lambda_j = 1$. The second coordinate of each x_j and therefore also of u is $\geq \alpha$. We have $\|u - x_j\| = t_1$ for all j.

Similarly, we can find $\{y_i\}_{i=1}^{s}$ in the closure of $K \cap C_2$ and $v = \sum \mu_i y_i$ with $\mu_i \geq 0$ and $\sum \mu_i = 1$ so that the second coordinate of v is $\leq -\alpha$ and so that $\|v - y_i\| = t_2 > \frac{1}{2}$ for all i. Note that $\|v - u\| \geq 2\alpha$. Since for every i and j

$$\|x_j - y_i\|^2 = \|x_j - u\|^2 + \|u - v\|^2 + \|v - y_i\|^2 +$$

$$+ 2\langle u - v, v - y_i \rangle + 2\langle x_j - u, u - v \rangle + 2\langle x_j - u, v - y_i \rangle \ ,$$

we get that

$$\sum_{i=1}^{s} \sum_{j=1}^{m} \lambda_j \mu_i \|x_j - y_i\|^2 = \|u - v\|^2 + t_1^2 + t_2^2 > 4\alpha^2 + \frac{1}{2} \ .$$

Since the diameter of K is at most 1 it follows that $\alpha < 2^{-3/2}$. Thus an upper estimate on d is obtained by taking $\alpha = 2^{-3/2}$. Coming back to the picture we get for this value of α

$$\sin \varphi = 2^{-3/2}(r^2 - 1/4)^{-1/2} \qquad , \qquad \cos \varphi = (r^2 - 3/8)^{1/2}(r^2 - 1/4)^{-1/2}$$

$$\sin \psi = (\rho^2 - 1/4)^{1/2}(r^2 - 1/4)^{-1/2} \quad , \quad \cos \psi = (r^2 - \rho^2)^{1/2}(r^2 - 1/4)^{-1/2} \quad .$$

Since $d = r \sin(\varphi + \psi)$ we deduce (5) . $\qquad\qquad \square$

Proof of the upper estimate in (1). Every set of diameter 1 in R^n is contained in a ball of radius $2^{-1/2}$. By cutting the ball into concentric shells of width $\varepsilon/4$ it follows from the form of (1) that it is enough to prove the possibility of covering sets of diameter 1 contained in S_r^{n-1} for some $1/2 \leq r \leq 1/2^{1/2}$. If $r \leq (3/8)^{1/2}$ then by the result of Rogers [R] S_r^{n-1} itself can be covered by $(\sqrt{3/2} + \varepsilon)^n$ balls of radius $\frac{1}{2}$. Hence we may assume from now on that $(3/8)^{1/2} < r \leq 1/2^{1/2}$.

We consider now the covers $\mathcal{A}_k = \{C_{k,m}\}$ of S_r^{n-1} by caps of diameter $2\rho_k$ as in Lemma 1 and denote the center of $C_{k,m}$ by $x_{k,m}$. Let $K \subset S_r^{n-1}$ be a subset of diameter 1 and put for $0 \leq k \leq k_0$,

$$D_k = \big\{x \in S_r^{n-1} \text{ , (the cap of diameter } 2\rho_k \text{ centered at } x) \cap K$$

$$\text{cannot be covered by 2 balls of radius } \tfrac{1}{2}\big\} \ .$$

By Lemma 2 the diameter of D_k is at most $2d(r, \rho_k)$ (see (5)). By a well known isoperimetric theorem, which goes back to Blaschke (and is proved by symmetrization on the sphere) the surface area of D_k relative to the area of S_r^{n-1} is at most $\big((d(r, \rho_k) + \varepsilon)/r\big)^n$ if $n > n(\varepsilon)$.

Let O^n denote the orthogonal group in R^n with normalized rotation invariant measure μ_n. For $U \in O^n$ and $0 \le k \le k_0$ put

$$N(k, U) = \left| \{ m \; ; \; U x_{k,m} \in D_k \} \right| .$$

Then, for $n \ge n(\varepsilon)$

$$\int_{O^n} N(k, U) d\mu_n(U) = M_k \cdot \text{area of } D_k \le M_k \big((d(r, \rho_k) + \varepsilon)/r \big)^n$$

and hence

$$\mu_n \big\{ U \; ; \; N(k, U) \ge M_k \big((1 + \varepsilon)(d(r, \rho_k) + \varepsilon)/r \big)^n \big\} \le (1 + \varepsilon)^{-n} .$$

Since the integer $k_0 = k_0(\varepsilon)$ appearing in Lemma 1 just depends on ε and not on n we get that for $n \ge n(\varepsilon)$ there is a $U_0 \in O^n$ such that for $0 \le k \le k_0$

$$N(k, U_0) \le M_k \big((1 + \varepsilon)(d(r, \rho_k) + \varepsilon)/r \big)^n \le \big((1 + \varepsilon)(r + \varepsilon)(d(r, \rho_k) + \varepsilon)/r \cdot \rho_k \big)^n . \qquad (6)$$

A direct calculation using formula (5) shows that for every $\sqrt{3/8} < r \le \sqrt{1/2}$ the maximal value of $d(r, \rho)/\rho$ (where ρ varies in $[\frac{1}{2}, r]$) is attained if $\rho^{-2} = r^{-2} + 4/3$ and the value of this maximum is $\sqrt{3/2}$ (and in particular independent of r). Hence it follows from (6) that

$$N(k, U_0) \le \big(\sqrt{3/2} + 10\varepsilon \big)^n , \qquad 0 \le k \le k_0 . \qquad (7)$$

Consider the covers $U_0 \mathcal{A}_k$ of S_r^{n-1}. The cover with the largest caps (i.e. for $k = k_0$) consists by (3) of at most $(10/9 + 2\varepsilon)^n \le \sqrt{3/2}^n$ caps. For a given cap $U_0 C_{k_0, m}$ in this cover we consider two alternatives. If the intersection of K with $U_0 C_{k_0, m}$ can be covered by two balls with radius $\frac{1}{2}$ we replace the cap by these two balls. If such two balls do not exist we replace it by $(1 + \varepsilon)^n$ caps $U_0 C_{k_0 - 1, j}$ as in (4). We continue in the same manner with the caps of level $k_0 - 1$ thus obtained. Either we replace the cap by two balls of radius $\frac{1}{2}$ which over its intersection with K or replace it by $(1 + \varepsilon)^n$ caps of level $k_0 - 2$. This process ends with $k = 0$, the caps on this level are themselves contained in balls of radius $\frac{1}{2}$. The number of caps of level $k < k_0$ obtained in the process is by (7) at most

$$(1 + \varepsilon)^n N(k + 1, U_0) \le \big((1 + \varepsilon)(\sqrt{3/2} + 10\varepsilon) \big)^n .$$

Hence the procedure above produces a cover of K by at most

$$2 k_0(\varepsilon) \big((1 + \varepsilon)(\sqrt{3/2} + 10\varepsilon) \big)^n$$

balls of radius $\frac{1}{2}$. □

Proof of the lower estimate in (1). The "critical" case in the proof of the upper estimate, i.e. subsets of $S^{n-1}_{\sqrt{3/8}}$, can be used also for obtaining a lower estimate in (1). Consider a subset K of $S^{n-1}_{\sqrt{3/8}}$ such that $x \neq y \in K$ implies $\|x - y\| \geq 2^{-1/2}$ and $\|x + y\| \geq 2^{-1/2}$ i.e. $|\langle x, y \rangle| \leq 1/8$. If this happens then $x \neq y \in K \Rightarrow \|x - y\| \leq 1$, i.e. K has a diameter of at most 1. Since $\|x - y\| \geq 2^{-1/2}$ for $x \neq y$ in K not more than n different point of K can be contained in a single ball of radius $\frac{1}{2}$ (by a result due to Rankin [Ra]). A simple volume estimate shows that K can be chosen to have at least $\left(\sqrt{9/8} - \varepsilon\right)^n$ points for $n \geq n(\varepsilon)$. This proves the lower estimate in (1). $\qquad\qquad\square$

In order to get a (slightly) better estimate we consider spheres of smaller radius $r = \frac{1}{2}(1 - t^2)^{1/2}$ with $t < 2^{-1/2}$. On such a sphere we consider again a set K so that $x, y \in K$, $x \neq y$ implies $\|x \pm y\| \geq t$. The diameter of such a K is again at most 1 and the volume estimate shows that K can be chosen to have at least $\left((1 + t^2)/2t - \varepsilon\right)^n$ points for $n \geq n(\varepsilon)$. From the results in [KL] it follows that a ball of radius $\frac{1}{2}$ contains at most $\left(\varphi(t) + \varepsilon\right)^n$ points whose mutual distance is at least t where

$$\varphi(t) = \left(\frac{1+u}{2u}\right)^{\frac{1+u}{2u}} \left(\frac{1-u}{2u}\right)^{\frac{u-1}{2u}} , \qquad u = u(t) = 2t(1 - t^2)^{1/2} .$$

The function $(1 + t^2)/2t\varphi(t)$ attains its maximum for $t \sim 0,69$ and the value at this point is $\sim 1,0645$. Hence

$$c_n \geq (1,0645)^n \qquad \text{for } n \geq n_0 . \tag{8}$$

References

[D]   L. Danzer, On the k-th diameter in E^d and a problem of Grünbaum, Proc. Colloquium on Convexity, Copenhagen, 1965, p. 41.

[G]   B. Grünbaum, Borsuk's problem and related question, Proc. Symp. Pure Math. A.M.S. 7 (1963), 271-284.

[KL]   G.A. Kabatiansky and V.I. Levenshtein, Bounds for packings on a sphere and in space, Problems in Information Transmission 14 (1978), 1-17.

[R]   C.A. Rogers, Covering a sphere with spheres, Mathematika 10 (1963), 157-164.

[Ra]   R.A. Rankin, On the closest packing of spheres in n dimensions, Ann. Math. 48 (1947), 1062-1081.

[S1]   O. Schramm, M.Sc. thesis, Hebrew University 1987.

[S2]   O. Schramm, Illuminating sets of constant width, Mathematika (1989).

CHARACTERIZATION OF
AFFINELY-ROTATION-INVARIANT LOG-CONCAVE MEASURES
BY SECTION-CENTROID LOCATION

M. Meyer

Equipe d'Analyse
Université Paris VI
Paris, France

S. Reisner

Department of Mathematics
University of Haifa
Haifa, Israel

§1. The starting point of this work is the theorem of Brunn (1889) which states: Let K be a convex body in $\mathbb{R}^3$ with the following property: the midpoints of every bundle of parallel chords of K lie in a plane. Then (and only then) K is an ellipsoid.

A generalization of this theorem was proved in [MR1] as follows: Let K be a convex body in $\mathbb{R}^n$ ($n \geq 2$) with the following property: for any $(n-1)$-dimensional subspace M of $\mathbb{R}^n$, the centroids of the sections $(x + M) \cap K$ are colinear (as x runs over all $x \in \mathbb{R}^n$ such that $\operatorname{int} K \cap (x + M) \neq \emptyset$). Then (and only then) K is an ellipsoid.

We reformulate this characterization of ellipsoids as follows. For $0 \neq a \in \mathbb{R}^n$ define

$$V(a) = \left| \{ x \in \mathbb{R}^n \ ; \ \langle x, a \rangle \geq 1 \} \cap K \right|$$

(where $|A|$ denotes the volume of the set A and $(\,,\,)$ denotes the usual scalar product in $\mathbb{R}^n$). Then (provided $\operatorname{int} K \cap \{ x \ ; \ \langle x, a \rangle = 1 \} \neq \emptyset$) the centroid of the section $K \cap \{ x \ ; \ \langle x, a \rangle = 1 \}$ is the point

$$x(a) = \nabla V(a) / \langle a, \nabla V(a) \rangle$$

(where ∇ denotes the gradient operator). Therefore the property assumed by K in the result mentioned above can be formulated as

$$(P) \quad \begin{cases} \text{for all } 0 \neq a \in \mathbb{R}^n \text{ and for all } s, t \in \mathbb{R} \setminus \{0\} \\ \nabla V(ta) \text{ is parallel to } \nabla V(sa) \,. \end{cases}$$

Now, let μ be a finite positive Borel measure in $\mathbb{R}^n$ and define

$$V(a) = \mu \left(\{ x \in \mathbb{R}^n \ ; \ \langle x, a \rangle \geq 1 \} \right) \,.$$

The result mentioned above leads us to consider the following problem.

Problem. Assuming that ∇V exists in some reasonable sense, does the property (P) imply (probably under additional conditions) that μ is *affinely-rotation-invariant?*

By the last adjective we mean the existence of an affine regular transformation L in $\mathbb{R}^n$ so that the measure $L\mu$ is *rotation invariant*, that is for every orthogonal transformation T of $\mathbb{R}^n$ holds

$$L\mu(TA) = L\mu(A)$$

$$(\text{equivalently } \mu(L^{-1}TLA) = \mu(A))$$

for all measurable sets $A \subset \mathbb{R}^n$.

In the particular case where K is a convex body and μ is defined by $\mu(A) = |A \cap K|$, the answer is 'yes' by the result mentioned before, L is in this case an affine transformation which maps the (a-posteriori) ellipsoid K onto a Euclidean ball centered at the origin.

In this paper we give a positive answer to the problem for a particular family of measures. Namely the symmetric and logconcave ones.

§2. *Log-Concave measures on $\mathbb{R}^n$.* A non-negative Borel measure μ on $\mathbb{R}^n$ is *logarithmically-concave* (log-concave) if $\mu(\theta A + (1 - \theta)B) \geq \mu(A)^\theta \mu(B)^{1-\theta}$ for all $0 \leq \theta \leq 1$ and A, B measurable sets (convex combination of sets here is in the sense of Minkowski). Brunn-Minkowski theorem is equivalent to the statement that Lebesgue measure in $\mathbb{R}^n$ is log-concave.

It was proved by Borell [Bo] and Prekopa [P] that if M is the minimal affine subspace containing the support of μ, then μ is log-concave if and only if there exists a log-concave function f, defined on M such that for all $A \subset M$

$$\mu(A) = \int_A f(x)dx$$

(dx being Lebesgue measure in M).

We shall call μ *non-degenerate* if no hyperplane contains the support of μ.

§3. Theorem 1. *Let μ be a finite, symmetric (say with respect to the origin) log-concave and non-degenerate measure on $\mathbb{R}^n$. The measure μ satisfies the property (P) if and only if it is affinely rotation invariant.*

Remarks.   a) The 'if' part in Theorem 1 is evident.

b) We conjecture that the pre-assumption of symmetry of μ is redundant. We know how to prove the theorem without assuming symmetry in the case that μ is not only log-concave but, moreover, that $\mu(\cdot)^{1/m}$ is a concave set function for some positive m.

We pass now to the proof of Theorem 1. First, we make the observation that, in the case of a log-concave measure, the meaning of the property (P) is very similar to that of the condition in the generalization of Brunn's theorem mentioned at the beginning. In fact, just as in [MR1] it can be checked that

$$x(a) = \nabla V(a)\big/\langle a, \nabla V(a)\rangle$$

is the centroid of the hyperplane

$$H(a) = \{x \ ; \ \langle x, a\rangle = 1\}$$

with respect to the mass density f restricted to $H(a)$, where f is the density of the log-concave measure μ.

That is

$$x(a) = \int_{\langle x,a\rangle=1} x f(x)dx \bigg/ \int_{\langle x,a\rangle=1} f(x)dx$$

(as long as $\int_{\langle x,a\rangle=1} f(x)dx > 0$. Notice also that the log-concavity of f, together with its integrability, imply that all the moments $\int |\langle x,y\rangle|^p f(x)dx$ exist for all y. Also, when we later use the term 'centroid' of a hyperplane with respect to μ – we mean this in the above sense).

The property (P) means that for every $(n-1)$-dimensional subspace M there exists a line $\ell(M)$ which contains all the centroids of hyperplanes parallel to M. It is easy to see that all these lines meet at one point – the centroid of $\mathbb{R}^n$ with respect to μ, which, by symmetry, must be the origin.

§4. *Floating bodies.*   For $0 < \alpha < \mu(\mathbb{R}^n)/2$ let $\mathcal{H}_\alpha$ be the family of all oriented hyperplanes H which satisfy $\mu(H^+) = \alpha$ (H^+ being the half space positive with respect to H).

Proposition 2. *If μ satisfies the assumptions of Theorem 1 (in particular – symmetry) then there exists a convex body K_α whose boundary is the envelope of the family $\mathcal{H}_\alpha$. That is, the family of supporting hyperplanes of K_α is identical with $\mathcal{H}_\alpha$ and K_α lies on the negative side* of every $H \in \mathcal{H}_\alpha$. Moreover, K_α is strictly convex and for every supporting hyperplane H of K_α, the contact point $H \cap K_\alpha$ is the centroid of H with respect to μ.

The convex bodies K_α are called 'floating bodies' of the measure μ.

In the case of a measure of the form $\mu(A) = |K \cap A|$ where K is a symmetric convex body in $\mathbf{R}^n$, Proposition 2 was proved in [MR2], independently, its main part – the convexity of K_α was proved (as yet unpublished) by K. Ball. We shall give later a proof of Proposition 2 in the more general, current situation. Meanwhile we proceed with the

§5. Proof of Theorem 1.

Let $0 < \alpha < \beta < \mu(\mathbf{R}^n)/2$. Consider the floating bodies $K_\beta \subset K_\alpha$. Let H_α and H_β be any pair of parallel supporting hyperplanes of K_α and K_β respectively. The contact points $H_\alpha \cap K_\alpha$ and $H_\beta \cap K_\beta$ which are the centroids of H_α and H_β, must, by the property (P) and the remark made at the end of §3, lie on a straight line which passes through the origin. Since this is true for every pair of parallel supporting hyperplanes of K_α and K_β, it follows easily that K_α and K_β are homothetical.

Fix now $0 < \alpha_0 < \mu(\mathbf{R}^n)/2$ and let $C = K_{\alpha_0}$. By the above discussion, there exists a decreasing positive function $\lambda(\alpha)$, defined on $\left(0, \mu(\mathbf{R}^n)/2\right)$ such that

$$K_\alpha = \lambda(\alpha)C$$

which implies

$$K_\alpha^* = \lambda(\alpha)^{-1}C^*$$

(K^* being the polar body of the symmetric convex body K).

We have for fixed $0 \neq y \in \mathbf{R}^n$

$$\int_{\mathbf{R}^n} |\langle x, y \rangle|^2 d\mu(x) = 2 \int_0^\infty \mu(\{x \; ; \; |\langle x, y \rangle| \geq t\}) t \, dt .$$

(The integral on the left hand side is finite by the remark on the moments of log-concave functions. From this, using Fubini's theorem, the existence of the integral on the right hand side and the equality follow).

Substituting in the last integral

$$t = \lambda(\alpha)\|y\|_{C^*}.$$

($\|\cdot\|_K$ denotes the norm induced by the symmetric convex body K) we get, using symmetry

$$\int_{\mathbf{R}^n} |\langle x, y \rangle|^2 d\mu(x) = -\|y\|_{C^*}^2 \cdot 4 \int_0^{\mu(\mathbf{R}^n)/2} \mu(\{x \; ; \; \langle x, \lambda(\alpha)^{-1}\|y\|_{C^*}^{-1}y \rangle \geq 1\})\lambda(\alpha)d\lambda(\alpha) .$$

But, by the definition of $\lambda(\alpha)$

$$\|\lambda(\alpha)^{-1}\|y\|_{C^*}^{-1}y\|_{K_\alpha^\circ} = 1 .$$

So, as $\{x \; ; \; \langle x, \lambda(\alpha)^{-1}\|y\|_{C^*}^{-1}y\rangle \geq 1\}$ is the set of points above a supporting hyperplane of K_α, we get

$$\mu(\{x \; ; \; \langle x, \lambda(\alpha)^{-1}\|y\|_{C^*}^{-1}y\rangle \geq 1\}) = \alpha$$

which, substituted in the above integral gives

$$\|y\|_{C^*} = \text{const} \cdot \left[\int_{\mathbb{R}^n} |\langle x, y\rangle|^2 d\mu(x)\right]^{1/2} .$$

This proves that C^* and C are ellipsoids.

Now we have

$$\mu(\{x \; ; \; \langle x, \lambda(\alpha)^{-1}y\rangle \geq 1\}) = \alpha \quad \text{if} \quad \|y\|_{C^*} = 1 .$$

Write this in the form

$$\mu(\{x \; ; \; \langle x, ty\rangle \geq 1\}) = \varphi(t) \quad \text{if} \quad \|y\|_{C^*} = 1$$

where φ is a function independent of the direction of y.

Let $L : \mathbb{R}^n \to \mathbb{R}^n$ be a linear transformation such that $L(C) = D_n$ (D_n is the Euclidean unit ball in $\mathbb{R}^n$ and $|\cdot|$ is the norm defined by it).
Then clearly

$$L\mu(\{w \; ; \; \langle w, tz\rangle \geq 1\}) = \varphi(t) \quad \text{if} \quad |z| = 1 .$$

It is now easy to use this last identity to conclude that the Fourier transform $\widehat{L\mu}$ of the (absolutely continuous) measure $L\mu$, as a function of $z \in \mathbb{R}^n$ depends on $|z|$ alone, from which it follows by the inversion formula that $L\mu$ is a rotation-invariant measure. This concludes the proof of Theorem 1.

§6. Proof of Proposition 2. We use an idea which was used by K. Ball in an unpublished proof of the convexity of floating bodies of symmetric convex bodies. This idea is to construct a norm in $\mathbb{R}^n$ and show that the unit ball of its dual norm is the floating body K_α.

For $a \in \mathbb{R}^n$ we define

$$B(a) = \{x \in \mathbb{R}^n \; ; \; |\langle x, a\rangle| \leq 1\}$$

$$W(a) = \mu(B(a))^{-1} .$$

We shall prove

Lemma 3. *The function* $W : \mathbf{R}^n \to \mathbf{R}^+$ *is convex (and even).*

Lemma 3 is the main ingredient in the proof of Proposition 2. Once we accept it we proceed as follows:

Assume for simplicity that $\mu(\mathbf{R}^n) = 1$. Given $0 < \alpha < 1/2$ define

$$\|a\| = \inf \left\{ \rho > 0 \; ; \; \mu \left(B \left(\frac{a}{\rho} \right) \right) \geq (1 - 2\alpha) \right\} .$$

The convexity of the function W implies that $\| \cdot \|$ is a norm on $\mathbf{R}^n$. Let $\| \cdot \|_*$ be its dual norm and K the unit ball of the norm $\| \cdot \|_*$. We have $\|x\|_* = 1$ if and only if $|\langle x, b \rangle| \leq 1$ for all b with $\|b\| = 1$ and there exists $a \in \mathbf{R}^n$ such that $\|a\| = \langle x, a \rangle = 1$, that is $\mu(B(a)) = 1 - 2\alpha$. So the hyperplane $H(a)$ is a supporting hyperplane of K at x and $V(a) = \mu(H(a)^+) = \alpha$. This proves that K is exactly the required floating body K_α. The strict convexity of K_α follows from differentiability of $V(a)$ just as in [MR2] (Theorem 3). The claim about the points of contact of supporting hyperplanes being the centroids of those hyperplanes, follows from the formula $x(a) = \nabla V(a) / \langle a, \nabla V(a) \rangle$ for those centroids $x(a)$, just as in [MR1] (Proposition 6).

Proof of Lemma 3. Let $a, b \in \mathbf{R}^n$, $a \neq b$, and suppose that $c = (a + b)/2 \neq 0$. Let

$$a' = \frac{\langle a, b \rangle a - |a|^2 b}{\left(|a|^2 |b|^2 - \langle a, b \rangle^2 \right)^{1/2}} , \quad b' = \frac{|b|^2 a - \langle a, b \rangle b}{\left(|a|^2 |b|^2 - \langle a, b \rangle^2 \right)^{1/2}} , \quad c' = \frac{a' + b'}{2} .$$

Then we have

$$|a'| = |a| , \quad |b'| = |b| , \quad |c'| = |c| , \quad \langle a, a' \rangle = \langle b, b' \rangle = \langle c, c' \rangle = 0$$

and

$$\langle a', b - a \rangle = \langle b', b - a \rangle = \langle c', b - a \rangle .$$

Let

$$H = \left\{ x \in \mathbf{R}^n \; ; \; \langle x, a - b \rangle = 0 \right\}$$

and

$$D = \left\{ x \in H \; ; \; |\langle x, a \rangle| \leq 1 \right\} = \left\{ x \in H \; ; \; |\langle x, b \rangle| \leq 1 \right\} = \left\{ x \in H \; ; \; |\langle x, c \rangle| \leq 1 \right\} .$$

Then we clearly have

$$B(a) = \bigcup_{t \in \mathbf{R}} (ta' + D) , \quad B(b) = \bigcup_{t \in \mathbf{R}} (tb' + D) , \quad B(c) = \bigcup_{t \in \mathbf{R}} (tc' + D) .$$

For a hyperplane J in $\mathbb{R}^n$ we denote by μ_J the Borell measure on J whose density is the density f of μ, restricted to J. By Fubini's theorem and using symmetry we have

$$\mu(B(a)) = \frac{2\langle a', a-b\rangle}{|a-b|} \int_0^\infty \mu_{ta'+H}(ta'+D)dt$$

$$\mu(B(b)) = \frac{2\langle b', a-b\rangle}{|a-b|} \int_0^\infty \mu_{tb'+H}(tb'+D)dt$$

$$\mu(B(c)) = \frac{2\langle c', a-b\rangle}{|a-b|} \int_0^\infty \mu_{tc'+H}(tc'+D)dt .$$

Since for $s,t > 0$, $\frac{t}{s+t}(sa'+D) + \frac{s}{s+t}(tb'+D) = \frac{2st}{s+t}c' + D$ it follows from log-concavity of μ (cf. e.g. [Bo] (Theorem 4.3)) that

$$\mu_{\frac{2st}{s+t}c'+H}\left(\frac{2st}{s+t}c'+D\right) \geq \mu_{sa'+H}(sa'+D)^{\frac{t}{s+t}}\mu_{tb'+H}(tb'+D)^{\frac{s}{s+t}} .$$

With the notation $h_y(t) = \mu_{ty+H}(ty+D)$ this is

$$h_{c'}\left(\frac{2st}{s+t}\right) \geq h_{a'}(s)^{\frac{t}{s+t}}h_{b'}(t)^{\frac{s}{s+t}} .$$

By a result of Ball [Ba] (cf. also [MP] Lemma 6) we get

$$\left(\int_0^\infty h_{c'}(t)dt\right)^{-1} \leq \frac{1}{2}\left[\left(\int_0^\infty h_{a'}(t)dt\right)^{-1} + \left(\int_0^\infty h_{b'}(t)dt\right)^{-1}\right] .$$

In view of the preceding relations, this gives

$$W\left(\frac{a+b}{2}\right) \leq \frac{1}{2}[W(a)+W(b)]$$

which completes the proof of Lemma 3.

§7.   *Lower dimensional characterizations* are easy to obtain as corollaries of Theorem 1. Thus we have

Theorem 4. Let μ be a finite symmetric, log-concave and non-degenerate measure on $\mathbb{R}^n$, such that for some (all) $1 \leq k \leq n-1$ and every k-dimensional subspace M, all the centroids (with respect to μ) of the flats $x + M$ lie in an $n-k$-dimensional flat. Then (and only then) μ is affinely rotation invariant.

References

[Ba]   K. Ball, Logarithmically concave functions and sections of convex sets in R^n, Studia Math. 88 (1988), 68-84.

[Bo]   C. Borell, Convex set functions in d-space, Period. Math. Hung. 6 (1975), 111-136.

[MP]   M. Meyer and A. Pajor, On the Blaschke-Santaló inequality, Arch. Math. 55 (1990), 82-93.

[MR1]   M. Meyer and S. Reisner, Characterizations of ellipsoids by section-centroid location, Geom. Ded. 31 (1989), 345-355.

[MR2]   M. Meyer and S. Reisner, A geometric property of the boundary of symmetric convex bodies and convexity of flotation surfaces, Geom. Ded. to appear.

[P]   A. Prekopa, On logarithmic concave measures and functions, Acta Sci. Math. 34 (1973), 335-343.

REMARKS ON MONTGOMERY'S CONJECTURES ON DIRICHLET SUMS

J. Bourgain

IHES
France

1. Introduction

This note deals mainly with two conjectures from [M] (p. 72,73) on the behaviour of general Dirichlet sums

$$\sum_{1}^{N} a_n n^{is} \; ; \qquad s = \sigma + it . \tag{1.1}$$

The first conjecture (9.1 in [M]) states in particular that if T is a 1-separated set of reals in the interval $[-T, T]$, then

$$\sum_{t \in \mathcal{T}} \left| \sum_{1}^{N} a_n n^{it} \right|^2 \le C_\varepsilon (N + |\mathcal{T}|^\varepsilon |\mathcal{T}|) \sum_{1}^{N} |a_n|^2 \tag{1.2}$$

for all $\varepsilon > 0$. Here $\{a_n\}$ are arbitrary scalars; $|\mathcal{T}|$ is the cardinality of the set $\mathcal{T}$. The distributional inequality derived from (1.2) yields immediately the moment inequality

$$\int_0^T \left| \sum_1^N a_n n^{it} \right|^{2\nu} dt \le (NT)^\varepsilon (T + N^\nu) \left(\sum |a_n|^2 \right)^\nu \tag{1.3}$$

for all $\nu \ge 1$. For $1 \le \nu \le 2$, (1.3) is a slightly weakened version of conjecture 9.2 of [M]. Taking $\nu = 1$, it is known that

$$\int_{T_0}^{T_0+T} \left| \sum_1^N a_n n^{it} \right|^2 dt \le C(T + N) \left(\sum_1^N |a_n|^2 \right) \tag{1.4}$$

(see [M], p.50) and hence (1.3) holds when ν is an integer. It is also shown in [M] that a statement such as (1.3) implies the density conjecture for the zeros of the Riemann zeta function $\zeta(s)$ in particular, i.e.

$$N(\sigma, T) \ll T^{2(1-\sigma)+\varepsilon} \tag{1.5}$$

(see [M], p. 100).

As will be observed here, both conjectures 1.2 and 1.3 are too strong and at least some restrictions on the coefficients $\{a_n \mid 1 \leq n \leq N\}$ have to be made. H. Montgomery proposed recently the following conjecture(∗) $(\log T \sim \log N)$

$$\sum_{t \in T} \left| \sum_{1}^{N} a_n n^{it} \right|^2 \ll N^\epsilon (N + |T|) N \tag{1.6}$$

assuming $|a_n| \leq 1$. As far as the zero-density estimates are concerned, (1.6) applies equally well. Conjecture (1.6) will also be discussed here. More precisely, we will derive from (1.6) certain function theoretic inequalities, implying in particular statements about the Hausdorff dimension of so-called Kakeya sets in higher dimensions: If (1.6) holds, then the following is true:

Any set in $\mathbf{R}^d$, $d \geq 2$, containing a translate of every line, has dimension d. (1.7)

This fact is true for $d = 2$ but open for $d \geq 3$. This last question belongs to an interesting problem area in geometric measure theory which is also of relevance to Fourier analysis.

Assuming the Lindelöf hypothesis, one has the estimate

$$\left| \sum_{1}^{N} n^{it} \right| \ll \frac{N}{|t|} + T^\epsilon N^{1/2} \tag{1.8}$$

implying (1.2) if $a_n = 1$, $1 \leq n \leq N$, $a_n = 0$ for $n > N$. On the other hand (see [M], p. 66), if we let

$$M(\alpha, T) = \max_{\substack{\sigma \geq \alpha, |t| \leq T \\ |s-1| \geq 1}} |\zeta(s)| \tag{1.9}$$

then one may write the following bound on the left member of (1.2)

$$\sum_{t \in T} \left| \sum_{N}^{2N} a_n n^{it} \right|^2 \ll (N + N^\alpha M(\alpha, 4T)|T|) \left(\sum_{N}^{2N} |a_n|^2 \right). \tag{1.10}$$

Thus, under the Lindelöf hypothesis, $|T|$ would carry a factor $(NT)^\epsilon \cdot N^{1/2}$ in (1.10). This inequality turns out to be essentially best possible for general ℓ^2-coefficients. Observe that (1.6) amounts to bound the Dirichlet operator(∗∗) $(n^{it})_{\substack{N \leq n \leq 2N \\ t \in T}}$ from $\ell^\infty_{[N, 2N]}$ to ℓ^2_T rather than from $\ell^2_{[N, 2N]}$ to ℓ^2_T. Unfortunately, the methods used in [M] or [B] are Hilbertian (estimates of $\|\mathcal{D}\|$ by $\|\mathcal{D}^*\mathcal{D}\|^{1/2}$) and it seems that new ideas are needed to make the distinction between the ℓ^∞ and ℓ^2-domain of $\mathcal{D}$.

(∗) Private communication
(∗∗) In the sense of [B], p. 32

The note ends with a proof that in (1.6) the factor N^ϵ has to appear in front and not only as a coefficient of $|T|$. In fact, we show that for $0 < \rho < 1$ there is a Dirichlet sum $S(s)$ of the form (1.1) with $|a_n| \le 1$ and a separated set $T \subset [0, N^2]$ for which

$$\sum_{t \in T} |S(it)|^2 > \exp(\log N)^\rho (N + |T| \cdot N^{\epsilon(\rho)}) N \tag{1.11}$$

where $\epsilon(\rho) > 0$. Improving this example seems a combinatorial problem.

I am grateful to P. Sarnak for some discussions on the subject matter of this note.

2. Short Sums

Fix N, $N < T < N^2$ and let k be a positive integer such that

$$T \cdot \frac{k^2}{N^2} = o(1) \tag{2.1}$$

(in fact, less than a suitable numerical constant).

For $0 \le r \le k$, $n = N + r$ and $|t| \le T$, write

$$n^{it} = N^{it} \left(1 + \frac{r}{N}\right)^{it} = N^{it} e^{i \frac{r}{N} t} e^{it[\log(1 + \frac{r}{N}) - \frac{r}{N}]}$$

where the last factor is of the form

$$1 + \sum_{\substack{j \ge 1 \\ j' \ge 2j}} c_{jj'} t^j \left(\frac{r}{N}\right)^{j'} = 1 + \sum_{\substack{j \ge 1 \\ j' \ge 0}} c'_{jj'} \left(\frac{t}{T}\right)^j \left(T \frac{k^2}{N^2}\right)^j \left(\frac{r}{k}\right)^{2j} \left(\frac{r}{N}\right)^{j'}.$$

Standard considerations permit then to estimate

$$\sum_{r=0}^{k} a_{N+r} e^{i \frac{r}{N} t} \tag{2.2}$$

by expressions of the form

$$\sum_{I} a_n n^{it}$$

where I runs over subintervals of $[N, N + k]$ (k was chosen such that on $[N, N + k]$ one may linearize $\log n$).

Take

$$\left. \begin{array}{ll} a_n = 1 & \text{if} \quad N \le n \le N + k \\ = 0 & \text{otherwise} \end{array} \right\} \tag{2.3}$$

and

$$T = \bigcup_{q=0,1,\ldots,\left[\frac{T}{2\pi N}\right]} (T_1 + 2\pi N q) \tag{2.4}$$

where T_1 is a 1-separated set in the interval $\left[0, \frac{N}{100k}\right]$. Thus

$$|T| \sim \frac{T}{k} \tag{2.5}$$

and

$$F(t) \equiv \sum_{r=0}^{k} e^{i\frac{r}{N}t} \tag{2.6}$$

is $\approx k$ for $t \in T$.

The left member of (1.2) is of order $k^2|T| \sim kT$ and hence

$$> c\left\{\left(\frac{T}{N}\right)N + k|T|\right\} \sum |a_n|^2 . \tag{2.7}$$

For a choice $T = N^{1+\delta}$, $0 < \delta < 1$, (2.7) equals

$$c\left(N^{1+\delta} + N^{\frac{1-\delta}{2}}|T|\right) \sum |a_n|^2 . \tag{2.8}$$

In particular, (1.2) fails and the optimal bound one may hope for is essentially $N(1 + T^\varepsilon N^{1/2}|T|) \sum |a_n|^2$, implied by the Lindelöf hypothesis.

Choosing $1 < \nu < 2$ and putting now

$$T = N^\nu \tag{2.9}$$

one also has

$$\int_0^T |F(t)|^{2\nu} dt \geq \int_\Omega |F(t)|^{2\nu} dt \geq T \cdot k^{2\nu-1} = T \cdot k^{\nu-1} \left(\sum |a_n|^2\right)^\nu \tag{2.10}$$

where

$$\Omega = \bigcup_{0 \leq q \leq \left[\frac{T}{2\pi N}\right]} \left[0, \frac{N}{100k}\right] . \tag{2.11}$$

Hence, also (1.3) fails for $1 < \nu < 2$.

Remark 2.12. For $\infty > p > 2$, call a 1-separated set $\Lambda \subset \mathbb{R}$ a Λ_p-set of constant $K \geq 1$, provided the inequality

$$\left(\int_0^1 \left| \sum_{\lambda \in \Lambda} a_\lambda e^{i\lambda t} \right|^p \right)^{1/p} \leq K \left(\sum_{\lambda \in \Lambda} |a_\lambda|^2 \right)^{1/2} \tag{2.13}$$

holds, for all scalar sequences $(a_\lambda)_{\lambda \in \Lambda}$. Letting $T = N^\nu$ in (1.3) and making a change of variable to normalize the interval of integration, (1.3) means that the set

$$\{T \cdot \log n \mid 1 \leq n < N\} \tag{2.14}$$

is a $\Lambda_{2\nu}$ set of constant N^ϵ. It is an elementary fact that such sets do not contain arithmetic progressions (or $\frac{1}{2}$-perturbations of arithmetic progressions) longer than $\sim N^{\epsilon/\nu-1}$. Now the set $\{T \cdot \log(N + r) \mid 0 \leq r \leq k\}$ where k is given by (2.1) is a perturbation of the progression $T \cdot \log N + \frac{T \cdot r}{N}$.

On the other hand, the inequality (cf. [M])

$$\int_0^{N^2} \left| \sum_1^N a_n n^{it} \right|^4 dt \leq CN^2 \sum_1^N |a_n(2)|^2 \tag{2.15}$$

derived from (1.4), yields that the set

$$\Lambda = \bigcup_{k=1}^\infty \{4^k \log p \mid 2^k < p < 2^{k+1} \, , \, p \text{ prime}\} \tag{2.16}$$

is a $\Lambda(4)$-set of bounded constant (and no $\Lambda(q)$-set for any $q > 4$). This is perhaps the simplest construction of such sets.

3. The Modified Conjecture (1.6)

This conjecture states a bound on the operator $\mathcal{D} = \{n^{it}\}_{\substack{1 \leq n \leq N \\ t \in T}}$ acting on $\ell^\infty_{[1,N]}$, thus

$$\sum_{t \in T} \left| \sum_1^N a_n n^{it} \right|^2 \ll N^\epsilon (N + |T|) N \cdot \max_{1 \leq n \leq N} |a_n|^2 \, . \tag{3.1}$$

Our purpose is to derive a function theoretic consequence of an estimate of the form

$$\sum_{t \in T} \left| \sum_1^N a_n n^{it} \right|^2 \leq (\alpha(N)N + \beta(N) \cdot |T|) \cdot N \cdot \max_{1 \leq n \leq N} |a_n|^2 \tag{3.2}$$

for T a 1-separated set in $[0, N^2]$. Here $\alpha(N), \beta(N)$ stand for slow growing functions of N. We will show the following.

Lemma 3.3. *Assume (3.2) holds. Let* $R, B \geq 1$ *and* $0 \leq f \leq 1$ *a function supported by the interval* $[0, B]$ *such that* $|f'| < R$. *Then*

$$\int_0^1 \sup_\tau \left[\sum_{b=1}^B f(\tau + bx) \right] dx \leq C \left(\alpha(R^2 B) + \beta(R^2 B) \cdot \int f \right) . \tag{3.4}$$

This inequality will be used to derive the dimension result of Kakeya sets mentioned above and to obtain (1.11). Let $N \leq T \leq N^2$ and k be as above, i.e.

$$Tk^2 \sim N^2 . \tag{3.5}$$

By (a_n) we always mean a 1-bounded sequence of scalars. Split $[N, 2N]$ into intervals I of length k. Clearly (3.2) implies the following "square function" inequality

$$\sum_{t \in T} \sum_I \left| \sum_{n \in I} a_n n^{it} \right|^2 < (\alpha \cdot N + \beta \cdot |T|) N . \tag{3.6}$$

Using the linear approximation to $\log n$ on the intervals I, the considerations in section 2 yield that

$$\sum_{t \in T} \sum_M^{2M} \left| \sum_{r=0}^{k-1} a_{mk+r} e^{i \frac{\tau}{mk} t} \right|^2 < C(\alpha N + \beta |T|) N \tag{3.7}$$

where

$$M = \frac{N}{k} . \tag{3.8}$$

For each $m = M, M+1, \ldots, 2M$, fix a point $t_m \in \mathbb{R}$ denoting

$$D(x) = \sum_0^{k-1} e^{ijx} \tag{3.9}$$

the Dirichlet kernel, a specification of the coefficients a_n in (3.7) gives

$$\sum_{t \in T} \sum_M^{2M} \left| D_k \left(\frac{t - t_m}{mk} \right) \right|^2 < C(\alpha N + \beta |T|) N . \tag{3.10}$$

Making a change of variable $t = kt'$, it follows that

$$\sum_{t' \in T'} \sum_M^{2M} \left| D_k \left(\frac{t' - t'_m}{m} \right) \right|^2 < C(\alpha N + \beta |T|) N . \tag{3.11}$$

where T' is an arbitrary $\frac{1}{k}$-separated set in $[0, \frac{T}{k}]$.

Consider a set $\mathcal{A}$ of 1-separated points in $\left[0, \frac{T}{k}\right]$ and put

$$T' = \sum_{t \in \mathcal{A}} \left(t + \left\{ \frac{j}{k} \mid 0 \le j < k \right\} \right) \tag{3.12}$$

which is a $\frac{1}{k}$-separated set of size

$$|T'| = k \cdot |\mathcal{A}| . \tag{3.13}$$

Since $k \ll M$, $D_k\left(\frac{t}{m}\right)$ is approximately constant on intervals of length 1. It follows from (3.12), (3.13) that

$$\sum_{t \in \mathcal{A}} \sum_{M}^{2M} \left| D_k\left(\frac{t - t_m}{m} \right) \right|^2 < C(\alpha \cdot M + \beta \cdot |\mathcal{A}|)N . \tag{3.14}$$

Consider the function χ_m, $0, 1$-values, defined as

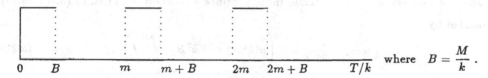

where $B = \dfrac{M}{k}$.

Because of the shape of $D_k\left(\frac{t}{m}\right)$, it follows from (3.14) that

$$\sum_{t \in \mathcal{A}} \sum_{M}^{2M} \chi_m(t - t_m) < C(\alpha M + \beta \cdot |\mathcal{A}|)B . \tag{3.15}$$

Consider the case where $\mathcal{A}$ is a union of intervals of length B in $\left[0, \frac{T}{k}\right] \cap \mathbb{Z}$. Thus

$$\mathcal{A} = \mathcal{A}' + [0, B] \tag{3.16}$$

where the points of $\mathcal{A}'$ are B-separated. Since

$$\left| (\mathcal{A}' + [0, B] - t_m) \cap (m\mathbb{Z} + [-B, B]) \right| \ge \frac{B}{2} \left| (\mathcal{A}' + [0, B] - t_m) \cap m\mathbb{Z} \right|$$

it follows from (3.15) that for such $\mathcal{A}$

$$\sum_{t \in \mathcal{A}} \sum_{M}^{2M} \mathcal{Y}_m(t - t_m) < C(\alpha M + \beta |\mathcal{A}|) \tag{3.17}$$

where $\mathcal{Y}_m$ denotes the indicator function of $m \cdot \mathbb{Z}$.

Rewrite (3.17) as

$$\sum_{M}^{2M} \sum_{j \in \mathbb{Z}} \chi_{\mathcal{A}} t_m + jm) < C(\alpha \cdot M + \beta \cdot |\mathcal{A}|) . \tag{3.18}$$

Let now f be the indicator function of a set in $[0, B]$ which is a union of intervals of length $\frac{1}{k} = \frac{B}{M}$.

One has

$$
\int_1^2 \sup_\tau \left(\sum_1^B f(\tau + bx) \right) dx = \sum_M^{2M} \int_{\left[\frac{m}{M}, \frac{m+1}{M}\right]} \sup_\tau \left(\sum_1^B f(\tau + bx) \right) dx
$$

$$
\approx \frac{1}{M} \sum_M^{2M} \sum_1^B f\left(\tau_m + b\frac{m}{M} \right)
$$

(3.19)

because of the assumption on f and for some choice of $\tau_m = \frac{t_m}{M}$. The set

$$
A = \left\{ t \in \mathbb{Z}, \quad 0 \le t \le M \cdot B \mid f\left(\frac{t}{M} \right) = 1 \right\}
$$

(3.20)

is a union of B-intervals and is contained in $\left[0, \frac{T}{k} \right]$, since $T \sim M \cdot B \cdot k$. Thus, by (3.18), (3.19) is bounded by

$$
C\left(\alpha + \beta \cdot \frac{|A|}{M} \right) \sim C\left(\alpha(k^2 B) + \beta(k^2 B) \cdot \int_0^B f \right) .
$$

(3.21)

Statement (3.4) is now clear, k corresponding to the derivative estimate R.

4. Consequences to the Hausdorff Dimension of Kakeya sets

We assume (1.6), i.e. (3.4) holds taking $\alpha(N) = \beta(N) = C_\varepsilon N^\varepsilon$, $\varepsilon > 0$ a fixed arbitrarily small number. Then the following statement holds

Let A be a subset of $\mathbb{R}^d$, $d \ge 2$ such that every line has a translate contained in A. Then $\dim A = d$.

(4.1)

Here dim stands for the Hausdorff dimension. As mentioned earlier, this is a fact for $d = 2$ (see [F] for details).

The main point consists of replacing (3.4) by its higher dimensional analogue. The argument is routine. This will permit us to derive a lower estimate for metrical entropy numbers.

Assume A is a union of subcubes of $[0, B]^d$ of size $\frac{1}{N}$. Assume that for every $x \in [0,1]^d$. There is some $\tau(x) \in \mathbb{R}^d$ such that

$$
\tau(x), \tau(x) + x, \ldots, \tau(x) + Bx \in A .
$$

Consider the standard partition of $\mathbb{R}_+^d$ into cubes of size $\frac{1}{N}$ which are labeled by a multi-index $(j_1, \ldots, j_d) \in \mathbb{Z}_+^d$. Put $R = N^d$. Define $Y_{(j_1, \ldots, j_d)}$ as the interval $(j_1 + j_2 N + \cdots +$

$j_d N^{d-1}) \frac{1}{R} + [0, \frac{1}{R}]$. For $j \in \mathbb{Z}$, let $\tilde{j}$ denote its residue mod N. Let $\mathcal{J}$ stands for the $(j_1, \ldots, j_d)$ corresponding to a cube contained in $\mathcal{A}$. Thus

$$|\mathcal{J}| = R \cdot \text{mes}(A) . \tag{4.2}$$

Consider the following subset of $[0,1]$

$$S = \bigcup_{(j_1, \ldots, j_d) \in \mathcal{J}} Y_{(\tilde{j}_1, \ldots, \tilde{j}_d)} . \tag{4.3}$$

Thus S is a union of $\frac{1}{R}$-intervals of measure

$$\text{mes}(S) \leq \text{mes}(A) . \tag{4.4}$$

Define the set $F \subset [0, B]$

$$F = S + \left\{ \frac{q_0}{N^d} + \frac{q_1}{N^{d-1}} + \frac{q_2}{N^{d-2}} + \cdots + q_d \mid q_i \in \mathbb{Z}, \ |q_i| < B \right\} \tag{4.5}$$

for which by (4.4)

$$\text{mes}(F) < (C \cdot B)^d \, \text{mes}(A) . \tag{4.6}$$

Assume $y \in Y_{(j_1, \ldots, j_d)} \subset [0,1]$. By hypothesis

$$\tau \left(\frac{j_1}{N}, \ldots, \frac{j_d}{N} \right) + b \left(\frac{j_1}{N}, \ldots, \frac{j_d}{N} \right) \in A \quad \text{for } 0 \leq b \leq B .$$

Thus for some $j'_1, \ldots, j'_d \in \mathbb{Z}_+$,

$$(j'_1 + bj_1, \ldots, j'_d + bj_d) \in \mathcal{J}, \qquad 0 \leq b \leq B .$$

Consequently, for $0 \leq b \leq B$

$$S \supset \frac{(j'_1 + bj_1)^\sim}{N^d} + \frac{(j'_2 + bj_2)^\sim}{N^{d-1}} + \cdots + \frac{(j'_d + bj_d)^\sim}{N} + \left[0, \frac{1}{R} \right] \tag{4.7}$$

and therefore, by construction of F

$$F \supset \left(\frac{j'_1}{N^d} + \frac{j'_2}{N^{d-1}} + \cdots + \frac{j'_d}{N} \right) + b \left(\frac{j_1}{N^d} + \frac{j_2}{N^{d-1}} + \cdots + \frac{j_d}{N} \right) + \left[\frac{-B}{R}, \frac{B}{R} \right] . \tag{4.8}$$

Since $y \in Y_{(j_1, \ldots, j_d)}$,

$$\left(\frac{j'_1}{N^d} + \frac{j'_2}{N^{d-1}} + \cdots + \frac{j'_d}{N} \right) + by \in F \quad \text{for } 0 \leq b \leq B . \tag{4.9}$$

Apply (3.4) with $f = \chi_F$. If $B > C_\varepsilon R^{3\varepsilon}$, it follows that

$$\int_0^B f > \frac{1}{C_\varepsilon} R^{-3\varepsilon} . \tag{4.10}$$

Hence, by (4.6)

$$\text{mes}(A) > (C_\varepsilon R^{3\varepsilon})^{-d} = R^{-\varepsilon'} . \tag{4.11}$$

From the preceding, a detailed deduction of (4.1) is now straightforward.

Remark. (4.1) is an open problem for $d = 3$. At the moment, I do know that A needs to have dimension $\geq \frac{7}{3}$.

5. Growth of the functions $\alpha(N), \beta(N)$ in (3.2)

Lemma 3.3 also permits us to get lower bounds on the numbers $\alpha(N), \beta(N)$. This is pursued next. The method is of combinatorial nature.

Let $S \subset [0,1] \times [0,1]$ be a union of $\left(\frac{1}{R} \times \frac{1}{R}\right)$-squares. Denote for $\varphi, \psi \in \mathbb{Z}_+$

$$P_{(\varphi,\psi)}(S) = \{\varphi x + \psi y \mid (x,y) \in S\} . \tag{5.1}$$

Let $1 \leq q \leq B$ and $\mathcal{B} \subset \{0, 1, \ldots, B\}$. If S satisfies

$$P_{(1,0)}(S) = [0,1] \tag{5.2}$$

then one has

$$|\mathcal{B}| \leq C\left\{\alpha(R^2 B) + \beta(R^2 B) \sum_{b \in \mathcal{B}} \operatorname{mes}\left(P_{(b,q)}(S)\right)\right\} . \tag{5.3}$$

Indeed, let $f = \chi_A$ where

$$A = \bigcup_{b \in \mathcal{B}} P_{(b,q)}(S) . \tag{5.4}$$

By (5.2) one may associate to each $x \in [0,1]$ a point τ' s.t. $(x, \tau') \in S$, hence $bx + q\tau' = bx + \tau \in A$ for all $b \in \mathcal{B}$. The left member of (3.4) is thus at least $|\beta|$ which yields inequality (5.3).

We now make the following construction. Let a and K be positive integers to be specified. Define

$$\xi = \left(1, -\frac{1}{a}\right) \qquad \eta = \left(\frac{1}{a}, 1\right) \tag{5.5}$$

$$S = \left\{\sum_{j=0}^{K-1} \frac{x_j \xi + y_n \eta}{a^{2j+1}} \mid x_j, y_j \in \{0, 1, \ldots, a-1\}\right\} \cup [0, a^{-2K}]^2 . \tag{5.6}$$

Thus

$$R = a^{2K} \tag{5.7}$$

and

$$P_{(\varphi,\psi)}(S) = \left\{\sum_{j=0}^{K-1} \frac{(\varphi x_j + \psi y_j)a + (\varphi y_j - \psi x_j)}{a^{2(j+1)}}\right\} + \left[0, (\varphi + \psi)a^{-2K}\right] .$$

In particular

$$P_{(1,0)}(S) = \left\{ \sum_{j=0}^{K-1} \frac{ax_j + y_j}{a^{2(j+1)}} \right\} + [0, a^{2K}] = [0, 1] \tag{5.8}$$

and for φ, ψ given by

$$\varphi = ca + d, \quad \psi = -c + da \, ; \qquad c, d \text{ positive integers } < a^\sigma \quad (\sigma < 1)$$

$$P_{(\varphi,\psi)}(S) \subset \left\{ (1 + a^2) \sum_{j=0}^{K-1} \frac{cx_j + dy_j}{a^{2(j+1)}} \right\} + [0, a^{-2(K-1)}] \tag{5.9}$$

implying

$$\text{mes } P_{(\varphi,\psi)}(S) \le (2a^{1+\sigma})^K a^{-2(K-1)} < R^{-\frac{1-\sigma}{3}} \tag{5.10}$$

if K is sufficiently large.

Observe also that the equation

$$\frac{ca + d}{-c + da} = \frac{c'a + d'}{-c' + d'a} \tag{5.11}$$

implies

$$(cd' - c'd)a^2 = c'd - cd'$$

hence

$$\frac{c}{d} = \frac{c'}{d'} \, . \tag{5.12}$$

We will let $d = 1$ and $|c| < a^\sigma$ vary.

Lemma 5.13. *Given a number U and $0 < \gamma < 1$, there is a set Ω of positive integers of size $\sim U$ and with a common multiple q such that*

$$|\Omega| > U^\gamma \tag{5.14}$$

and

$$q < \exp(\log U)^{\frac{1}{1-\gamma}} \, . \tag{5.15}$$

Proof: Choose a prime $p = p_1$ such that

$$\log U \sim p^{1-\gamma} \tag{5.16}$$

and r with

$$10^r p^r \le U < 10^{r+1} p^{r+1} \, . \tag{5.17}$$

Let $q_0 = p_1 p_2 \cdots p_s$ be the product of $s \sim \frac{p}{\log p}$ consecutive primes and Ω_0 the set of all simple products x of r primes among $p_1, \ldots, p_s$. Each such x is less than U, by (5.17), and may thus be multiplied with a divisor of $\prod_{p' < p_s} p'$ to obtain $x' \sim U$. The set Ω of those numbers admits $q = \prod_{p' < p_s} p'$ as a common multiple and

$$|\Omega| = |\Omega_0| = \binom{s}{r} . \tag{5.18}$$

It follows from (5.16),(5.17) that $r \log p \sim p^{1-\gamma}$, hence $\frac{s}{r} \sim p^{\gamma}$ and (5.14). Also (5.15) holds by (5.16).

□

Assume $\Omega \subset [U, 10U]$. Split the interval $[U, 10U]$ into $\sim U^{1-\sigma}$ intervals of the form $[a - a^{\sigma}, a]$ ($\sigma < 1$). For $0 < \gamma < 1$ and

$$\sigma > 1 - \gamma \tag{5.19}$$

one may find such an interval $[a - a^{\sigma}, a]$ containing a subset Ω' of Ω of size at least

$$|\Omega'| > U^{\gamma - 1 + \sigma} . \tag{5.20}$$

Defining S as in (5.6), and putting

$$\mathcal{B} = \left\{ (ca + 1) \frac{q}{a - c} \;\middle|\; a - c \in \Omega' \right\} \tag{5.21}$$

it follows from (5.3),(5.7),(5.20)

$$U^{\gamma - 1 + \sigma} \leq \alpha(a^{4K+2}q) + \beta(a^{4K+2}q) \sum_{a - c \in \Omega'} \frac{q}{a - c} \, \mathrm{mes} \, P_{(ca+1, a-c)}(S) . \tag{5.22}$$

By (5.10),(5.15), one deduces from (5.22)

$$U^{\gamma - 1 + \sigma} \leq \alpha \left((20U)^{4K} \exp(\log U)^{\frac{1}{1 - \gamma}} \right) + \beta \left((20U)^{4K} \exp(\log U)^{\frac{1}{1 - \gamma}} \right) \cdot \exp(\log U)^{\frac{1}{1 - \gamma}} \cdot U^{-\frac{1 - \sigma}{2} K} . \tag{5.23}$$

Choosing

$$K \sim \frac{1}{1 - \sigma} (\log U)^{\frac{\gamma}{1 - \gamma}} \tag{5.24}$$

one gets

$$U^{\gamma - 1 + \sigma} \leq \alpha \left(\exp \frac{1}{1 - \sigma} (\log U)^{\frac{1}{1 - \gamma}} \right) + \beta \left(\exp \frac{1}{1 - \sigma} (\log U)^{\frac{1}{1 - \gamma}} \right) \cdot \exp \left(-(\log U^{\frac{1}{1 - \gamma}}) \right) . \tag{5.25}$$

Putting

$$N \sim \exp \frac{1}{1 - \sigma} (\log U)^{\frac{1}{1 - \gamma}} \tag{5.26}$$

one gets

$$\exp\left[(\gamma - 1 + \sigma)(1 - \sigma)(\log N)^{1-\gamma}\right] \leq \alpha(N) + \beta(N)N^{-(1-\sigma)} \, . \qquad (5.27)$$

Here γ, σ are subject to (5.19). Thus, one sees that for each $0 < \rho < 1$ there is $\varepsilon(\rho) > 0$ such that either $\alpha(N) > \exp(\log N)^{\rho}$ or $\beta(N) > N^{\varepsilon(\rho)}$. This corresponds to (1.11).

References

[B] E. Bombieri, Le grand crible dans la théorie analytique des nombres. Astérisque 18 (1974).

[F] K. Falconer, The Geometry of Fractal Sets, Cambridge UP, 1985.

[M] H. Montgomery, Topics in multiplicative number theory. Springer Lecture Notes in Math. 227 (1971).

ON THE DEPENDENCE ON ε IN A THEOREM OF
J. BOURGAIN, J. LINDENSTRAUSS AND V.D. MILMAN

M. Schmuckenschläger

Institut für Mathematik
Johannes Kepler Universität Linz
A-4040 Linz, Austria

In [BLM1] it was shown that for any convex compact set K in $\mathbb{R}^n$ there exist orthogonal transformations $(U_j)_{j=1}^N$ on $\mathbb{R}^n$, with

$$N \leq cn\varepsilon^{-2} \log \frac{1}{\varepsilon}$$

so that the Minkowski sum of the sets

$$\left(\frac{1}{N} U_j(K) \right)_{j=1}^N$$

is up to ε a Euclidean ball of radius

$$r = \int\limits_{n-1} \|x\|_* d\sigma_n(x)$$

($\| \ \|_*$ denotes the norm whose unit ball is the polar to $(K - K)$ and σ_n is the unique rotation invariant probability measure S^{n-1}).

The purpose of this note is to prove the same statement without the $\log \frac{1}{\varepsilon}$ factor. We use a variation of the arguments of Schechtman [S], who proved that for any norm $\| \cdot \|$ on $\mathbb{R}^n$ there exists an operator

$$T : l_n^2 \xrightarrow{\text{onto}} Y \subseteq \left(\mathbb{R}^N, \| \cdot \| \right) \qquad \text{with}$$

$$n \geq c\varepsilon^2 \left(\frac{E}{\sigma} \right)^2 \qquad \text{and}$$

$$\|T\|\|T^{-1}\| \leq 1 + \varepsilon$$

(E denotes the expectation of $\sim \sqrt{N}\| \cdot \|$ on S^{N-1}, σ is the norm of the identity map from l_N^2 to $\left(\mathbb{R}^N, \| \cdot \| \right)$ and c is an absolute constant). Schechtman considered a random operator

$$R : l_N^2 \to \left(\mathbb{R}^N, \| \cdot \| \right)$$

whose matrix representation with respect to the canonical bases of $\mathbf{R}^n$ and $\mathbf{R}^N$ respectively has independent standard Gaussian entries. In probabilistic terms the point of his proof was to show that the process

$$\left(\|T(w)x\| - E\right)_{x \in S^{n-1}}$$

is subgaussian, i.e.

$$P\left(\left|\|Tx\| - \|Ty\|\right| > t\right) \le 2 \exp -c \left(\frac{t}{\sigma\|x - y\|_2}\right)^2, \qquad x, y \in S^{n-1}, \, t > 0$$

or, equivalently (see [MaP])

$$E \exp \left(\lambda(\|Tx\| - \|Ty\|)\right) \le \varepsilon\left(K\lambda^2\sigma^2\|x - y\|_2^2\right), \qquad \forall \lambda \ge 0, \, x, y \in S^{n-1}.$$

We start by recalling the basic tool from the empirical distribution method namely the so-called Bernstein inequality. Before stating the result, we recall a definition [MaP].

Definition. Let X be a (real valued) random variable. For $\alpha \ge 1$ we define the N_α-norm of X by

$$N_\alpha(X) := \inf\left\{\lambda > 0 : E \exp\left(\frac{|X|}{\lambda}\right)^\alpha \le 2\right\}.$$

This is the Orlicz norm associated with the function $\psi_\alpha(t) = \exp(t^\alpha) - 1$.

Lemma 1. Let $(X_j)_{j=1}^N$ be independent random variables with expectation zero. Assume that $N_2(X_j) \le A$. Then

$$P\left(\left|\sum_{j=1}^N X_j\right| > Nt\right) \le 2 \exp -\frac{N}{8}\left(\frac{t}{A}\right)^2.$$

For the random operator T considered by Schechtman we have the estimate

$$N_2\left(\|Tx\| - \|Ty\|\right) \le c\sigma\|x - y\|_2, \qquad \forall x, y \in S^{n-1}.$$

Indeed, for $x, y \in S^{n-1}$ define

$$z = \tfrac{1}{2}(x + y) \qquad u = \tfrac{1}{2}(x - y).$$

Then $u \perp z$, $x = z + u$, $y = z - u$ and therefore Tz and Tu are independent. Hence, by the result of Pisier (see (MS])

$$E \exp \left(\lambda(\|Tx\| - \|Ty\|)\right) = EE \exp \left(\lambda(\|Tz + Tu\| - \|Tz - Tu\|)\right)$$

$$= \int E \exp \left(\lambda(\|\bar{z} + Tu\| - \|\bar{z} - Tu\|)\right) dP_{Tz}(\bar{z})$$

$$\le \int \exp \left(\frac{\lambda^2\pi^2}{8}(2\sigma\|u\|_2)^2\right) dP_{Tz}(\bar{z})$$

$$= \exp \left(c'\lambda^2\sigma^2\|x - y\|_2^2\right).$$

Thus the process $\left(\|Tx\|\right)_{x \in S^{n-1}}$ is subgaussian and

$$N_2\left(\|Tx\| - \|Ty\|\right) \le c\sigma \|x - y\|_2 \ .$$

Remark. This proof is essentially due to Schechtman [S]. However, we did not use conditional probabilities. Since the random variables Tu and Tz are independent there is no need for conditioning.

Proposition 1. Let $(T_j)_{j=1}^{N}$ be a sequence of independent random operators on $\mathbb{R}^n$ such that

$$(i) \quad E := E\|T_j x\| \qquad\qquad \forall j \le N \ , \forall x \in S^{n-1}$$

$$(ii) \quad N_2\left(\|T_j x\|\right) \le A_0 \qquad \forall j \le N \ , \ \forall x \in S^{n-1}$$

Then for $N \ge cn\varepsilon^{-2}\left(\frac{A_0}{E}\right)^2$ we have for some $w \in \Omega$ and all $x \in \mathbb{R}^n$

$$(1 - \varepsilon)E\|x\|_2 \le \frac{1}{N}\sum_{j=1}^{N} \|T_j(w)x\| \le (1 + \varepsilon)E\|x\|_2 \ .$$

Proof: Since for all $j \le N$, $x, y \in S^{n-1}$ $\|T_j x\| - \|T_j y\|$ has mean zero and

$$N_2\left(\|T_j x\| - \|T_j y\|\right) \le \|x - y\|_2 N_2\left(\left\|T_j\left(\frac{x - y}{\|x - y\|_2}\right)\right\|\right)$$

$$\le A_0 \|x - y\|_2$$

the Bernstein inequality yields

$$P\left(\frac{1}{N}\left|\sum \|T_j x\| - \|T_j y\|\right| > t\right) \le 2\exp - \frac{N}{8}\left(\frac{t}{A_0\|x - y\|_2}\right)^2 \ .$$

We now proceed as in Schechtman's proof of the Dvoretzky Theorem:

Let F_k be a 2^{-k}-net in S^{n-1}. Then

$$|F_k| \le (1 + 2^{k+1})^n \le (3 \cdot 2^k)^n$$

For $k \in \mathbb{N}$ and $x \in S^{n-1}$ there exist $x_k \in F_k$ such that $\|x - x_k\|_2 \le 2^{-k}$. Hence

$$\|x_k - x_{k-1}\|_2 \le \|x_k - x\|_2 + \|x_{k-1} - x\|_2 \le 3 \cdot 2^{-k} \ .$$

For $\delta_k \ge 0$ with $\sum_{k=1}^{\infty} \delta_k = 1$ and $w \in \Omega$ with

$$\left\|\frac{1}{N}\sum_{j=1}^{N} \|T_j(w)x\| - E\right| > \varepsilon E$$

we get that for some $k \geq 2$

$$\left| \frac{1}{N} \left(\sum_{j=1}^{N} \|T_j(w)x_k\| - \|T_j(w)x_{k-1}\| \right) \right| > \varepsilon \delta_k E$$

or

$$\left| \frac{1}{N} \sum_{j=1}^{N} \|T_j(w)x_1\| - E \right| > \varepsilon \delta_1 E \; .$$

Therefore

$$P\left(\exists x \in S^{n-1} : \left| \frac{1}{N} \sum_j \|T_j x\| - E \right| > \varepsilon E \right)$$

$$\leq P\left(\exists k \leq 2 \, , \; x_k \in F_k \, , \; x_{k-1} \in F_{k-1} : \|x_k - x_{k-1}\| \leq 3 \cdot 2^{-k} \quad \text{and} \right.$$

$$\left. \left| \frac{1}{N} \left(\sum_j \|T_j x_k\| - \|T_j x_{k-1}\| \right) \right| > \varepsilon \delta_k E \right)$$

$$+ P\left(\exists x \in F_1 : \left| \frac{1}{N} \sum_j \|T_j x_1\| - E \right| > \varepsilon \delta_1 E \right)$$

$$\leq \sum_{k=2}^{\infty} |F_k||F_{k-1}| P\left(\frac{1}{N} \left| \sum \|T_j x_k\| - \|T_j x_{k-1}\| \right| > \varepsilon \delta_k E \right)$$

$$+ |F_1| P\left(\left| \frac{1}{N} \sum \|T_j x_1\| - E \right| > \varepsilon \delta_1 E \right)$$

$$\leq 2 \sum_{k=2}^{\infty} (3 \cdot 2^k)^{2n} \exp -\frac{N}{8} \left(\frac{\varepsilon \delta_k E}{3A \cdot 2^{-k}} \right)^2 + 2(3 \cdot 2^1)^{2n} \exp -\frac{N}{8} \left(\frac{\varepsilon \delta_1 E}{A_0} \right)$$

$$\leq 2 \sum_{k=1}^{\infty} (3 \cdot 2^k)^{2n} \exp -\frac{N}{8} \left(\frac{\varepsilon \delta_k E}{3A_0 2^{-k}} \right)^2$$

$$\leq 2 \sum_{k=1}^{\infty} \exp \left(2nk \log 6 - \frac{1}{72} N \varepsilon^2 \delta_k^2 2^{2k} \left(\frac{E}{A_0} \right)^2 \right)$$

For $\delta_k = c_1 \sqrt{k} 2^{-k}$ we get that the last expression is

$$2 \sum_{k=1}^{\infty} \exp k \left(2n \log 6 - c_2 \varepsilon^2 N \left(\frac{E}{A_0} \right)^2 \right)$$

which is smaller than 1 for

$$n \geq cn\varepsilon^{-2} \left(\frac{A_0}{E} \right)^2 .$$

□

If we choose orthogonal operators U_j which are uniformly distributed on the orthogonal group $O(n)$ for the T_j, then these random operators clearly satisfy the hypothesis (i) of the preceding proposition and, by a result of Marcus and Pisier [MP],

$$N_2(\|Ux\|) \leq cE\|Ux\| = cE .$$

We now apply the proposition to the polar of $(K - K)$ and obtain the following result:

Theorem 1. *Let K be a compact convex set in $\mathbb{R}^n$. Then there exist orthogonal transformations $(U_j)_{j=1}^N$ with*

$$N \leq cn\varepsilon^{-2}$$

so that

$$B_n(1 - \varepsilon)E^* \subset \frac{1}{N}\sum_{j=1}^N U_j(K) \subseteq (1+\varepsilon)E^* B_n$$

where B_n denotes the unit ball of ℓ_n^2 and $E^* = \int\limits_{S^{n-1}} \|x\|_* d\sigma_n(x)$.

If K is a convex symmetric body in $\mathbb{R}^n$ so that the Euclidean ball B_n is the ellipsoid of minimal volume containing K, then, as was shown in [BLM2], we can get a better estimate for N

$$N \leq c\frac{n}{\log n}\varepsilon^{-2}\log\frac{1}{\varepsilon} .$$

Even in this case, the factor $\log\frac{1}{\varepsilon}$ can be dropped, if we use the complete proof of Schechtman.

In view of the proof of Proposition 1 it suffices to prove the following

Lemma. *Let U be a random operator which is uniformly distributed on $O(n)$. Let σ be the norm of the identity map from l_n^2 to $(\mathbb{R}^n, \|\cdot\|)$. Then for all $x, y \in S^{n-1}$*

$$P(|\|Ux\| - \|Uy\|| > t) \leq \exp\left(-cn\left(\frac{t}{\sigma\|x-y\|_2}\right)^2\right) .$$

Proof: Define

$$z = \tfrac{1}{2}(x + u) , \qquad u = \tfrac{1}{2}(x - y) , \qquad v = -u .$$

Then

$$u \perp z , \quad v \perp z , \quad x = u + z \quad \text{and} \quad \|u\|_2 = \|v\|_2 = \tfrac{1}{2}\|x - y\|_2 .$$

Conditioned on $Uz = \bar{z}$ the random variable Uu and Uv are uniformly distributed on

$$S^{n-1} \cap (\bar{z} + [\bar{z}]^\perp) .$$

If we define

$$F : O(n-1) \to \mathbf{R}$$

by the equation

$$F(V) = \|\bar{z} + Vu\| - \|\bar{z} + Vv\| \,,$$

then the Lipschitz constant $\mathrm{Lip}(F)$ with respect to the Hilbert-Schmidt metric on $O(n-1)$ is at most

$$\sigma\big(\|u\|_2 + \|v\|_2\big) = \sigma\|x - y\|_2 \,.$$

It now follows from the fact that $F(V)$ has expectation zero and the concentration of measure property of $O(n)$ ([MS], [GM]) (since only one vector is involved, it is actually the concentration phenomenon on the sphere) that

$$P\big(\big|\|Ux\| - \|Uy\|\big| > t | Uz = \bar{z}\big) = \mu\big(|F(V)| > t\big)$$

$$\leq \exp{-cn}\left(\frac{t}{\sigma\|x - y\|_2}\right)^2$$

where μ denotes Haar measure on $O(n-1)$.

Since

$$P\big(\big|\|Ux\| - \|Uy\|\big| > t\big) = \int P\big(\big|\|Ux\| - \|Uy\|\big| > t \mid Uz = \bar{z}\big) dP_{U_z}(\bar{z})$$

the result follows. $\qquad\qquad\qquad\qquad\qquad\qquad\qquad\qquad\qquad\qquad\qquad\qquad\square$

Theorem 2. *Let K be a convex symmetric body in $\mathbf{R}^n$ such that the unit ball B_n of l_n^2 is the ellipsoid of minimal volume containing K. Then there are orthogonal transformations $(U_j)_{j=1}^N$ with*

$$N \leq c\frac{n}{\log n}\varepsilon^{-2}$$

so that

$$(1 - \varepsilon)E^*B_n \subseteq \frac{1}{N}\sum_{j=1}^N U_j(K) \subseteq (1 + \varepsilon)E^*B_n \,.$$

Proof: The assumption on K implies that B_n is the ellipsoid of maximal volume contained in the polar K^0 of K. Replacing the estimates on the probabilities in the proof of Proposition 1 by those of the previous lemma we obtain the required conclusion for

$$N \geq cn\varepsilon^{-2}\left(\frac{1}{\sqrt{n}E^*}\right)^2 \,.$$

In [FLM] it was shown that under the above conditions on K^0 we have

$$E^* \geq c \left(\frac{\log n}{n} \right)^{1/2} .$$

This gives the desired estimate for N. □

If we choose the operator

$$Rx := \sum_{i=1}^{n} r_i(t)x_i e_i \qquad (x = \sum_{i=1}^{n} x_i e_i)$$

and if we denote by $(R_j)_{j=1}^{N}$ independent copies of R, then these operators do not satisfy (i). So we define [see BLM2] a new norm on $\mathbf{R}^n$

$$\|x\|_0 := E\|Rx\| .$$

Let C_0 be the unit ball with respect to the norm $\| \cdot \|_0$. Then the process

$$(Rx)_{x \in \partial C_0}$$

satisfies the hypothesis (i) of the proposition (S^{n-1} replaced by ∂C_0). On the other hand the Kahane inequality gives

$$N_2(\|Rx\|) \leq cE\|Rx\| = c \qquad \forall x \in \partial C_0 .$$

Therefore we obtain Proposition 4 of [BLM2] with

$$N \leq cn\varepsilon^{-2} .$$

References

[BL]   J. Bourgain and J. Lindenstrauss, Distributions of points on spheres and approximation by zonotopes, Israel J. of Math. 64 (1989).

[BLM1]   J. Bourgain, J. Lindenstrauss and V.D. Milman, Approximation of zonoids by zonotopes, Acta Math. 162, 1989.

[BLM2]   J. Bourgain, J. Lindenstrauss and V.D. Milman, Minkowski sums and symmetrizations, Springer Lecture Notes 1317, GAFA 1986-1987.

[FLM]   T. Figiel, J. Lindenstrauss and V.D. Milman, The dimension of almost spherical sections of convex bodies, Acta Math. 139, 1977.

[GM1]   M. Gromov and V.D. Milman, An application of the isoperimetric inequality, Amer. J. Math. 105, 1983.

[MS]   V.D. Milman and G. Schechtman, Asymptotic theory of finite dimensional normed spaces, Springer Lecture Notes 1200, 1986.

[MaP]   M.B. Marcus and G. Pisier, Random Fourier series with applications to harmonic analysis, Ann. Math. Studies 101, Princeton 1981.

[S]   G. Schechtman, A remark concerning the dependence on ε in Dvoretzky's theorem, Springer Lecture Notes 1376, GAFA 1987-88.

ANOTHER REMARK ON

THE VOLUME OF THE INTERSECTION OF TWO L_p^n BALLS

G. Schechtman*   and   M. Schmuckenschläger

Department of Theoretical Mathematics   Institut für Mathematik
The Weizmann Institute of Science   J. Kepler Universität
Rehovot, Israel   Linz, Austria

Let B_p^n be the multiple of the ℓ_p^n ball which has volume one; i.e.,

$$B_p^n = \left\{ (x_1, x_2, \ldots, x_n) \; ; \; \left(\sum_{i=1}^{n} |x_i|^p\right)^{1/p} \leq \frac{\Gamma(1 + n/p)^{1/n}}{2\Gamma(1 + 1/p)} \right\}$$

if $0 < p < \infty$ (see [M] p. 121 or [S] p. 25-26 for the fact that $\mathrm{Vol}(B_p^n) = 1$), and

$$B_\infty^n = \left\{ (x_1, x_2, \ldots, x_n) \; ; \; \max_{1 \leq i \leq n} |x_i| \leq \frac{1}{2} \right\}.$$

Using the technique developed in [SZ], We investigate here the asymptotic behaviour of the volume of the intersection of multiples of any two of the balls B_p^n. More precisely, put

$$A_{p,q} = e^{1/p - 1/q} \Gamma(1 + 1/p)^{1 + 1/q} p^{1/q} \Gamma(1 + 1/q)^{-1} \Gamma\left(\frac{q+1}{p}\right)^{-1/q} q^{-1/q}, \quad 0 < p, q < \infty \quad (1)$$

$$A_{\infty,q} = (q+1)^{1/q} e^{-1/q} q^{-1/q} \Gamma(1 + 1/q)^{-1}, \qquad\qquad\qquad 0 < q < \infty \quad (2)$$

Theorem. For all $0 < p \leq \infty$, $0 < q < \infty$, as $n \to \infty$,

$$\mathrm{Vol}(B_p^n \cap tB_q^n) \to 0, \quad if \quad tA_{p,q} < 1$$

and

$$\mathrm{Vol}(B_p^n \cap tB_q^n) \to 1, \quad if \quad tA_{p,q} > 1.$$

It follows from the Theorem that $A_{p,q} \leq 1$. We also show

* Supported in part by the Glikson Foundation

Proposition. _a_ For all $0 < p < q < \infty$, $A_{p,q} < 1$.

_b_ For all $0 < q < \infty$, $A_{\infty,q} < 1$.

Thus getting as a corollary,

Corollary. _For all $0 < p, q \le \infty$ with $p \ne q$,_

$$\mathrm{Vol}(B_p^n \cap B_q^n) \to 0.$$

The last corollary, in the three special cases $\{p, q\} \subset \{1, 2, \infty\}$, answers a question of J. M. Wills (no. 21 in [M]) the case $p = 1$, $q = \infty$ of which has previously been solved by B. Weißbach [W].

For the proof of the Theorem we recall some of the notations and one of the Lemmas of [SZ] with some adaptations:

Put, for $0 < p < \infty$

$$r_p = \frac{\Gamma(1 + n/p)^{1/n}}{2\Gamma(1 + 1/p)}$$

and

$$\Delta_p = \{(t_1, t_2, \ldots, t_n) \, ; \, t_i \ge 0, \sum t_i^p = r_p^p\} = \partial B_p^n \cap \mathbb{R}^{+n} \, ,$$

Then for $A \subset B_p^n \cap \mathbb{R}^{+n}$

$$\mathrm{Vol}(A) = 2^{-n} n \int_0^1 r^{n-1} \mu_p(r^{-1}A) dr$$

where μ_p is the Borel probability measure on Δ_p given by

$$\mu_p(A) = \lim_{\epsilon \to 0} \frac{\mathrm{Vol}(\{(x_1, \ldots, x_n) \in \mathbb{R}^+ A \, ; (r_p - \epsilon)^p < \sum t_i^p < (r_p + \epsilon)^p\})}{\mathrm{Vol}(\{(x_1, \ldots, x_n) \in \mathbb{R}^{+n} \, ; (r_p - \epsilon)^p < \sum t_i^p < (r_p + \epsilon)^p\})}.$$

It follows from Lemma 1 in [SZ] that if $X_1, \ldots, X_n$ are independent random variables all with density function

$$p\Gamma(1/p)^{-1} e^{-t^p} \, , t > 0,$$

then

$$r_p(\sum_{i=1}^n |X_i|^p)^{-1/p}(X_1, \ldots, X_n)$$

induces the measure μ_p on Δ_p.

Proof of the Theorem. Fix $0 < p, q < \infty$ and $0 < t < \infty$. Then by the discussion above,

$$\mathrm{Vol}(B_p^n \cap tB_q^n) = \mathrm{Vol}(\{(x_1, \ldots, x_n) \in B_p^n \ ; \ (\sum_{i=1}^n |x_i|^q)^{1/q} \le tr_q\})$$

$$= n \int_0^1 r^{n-1} \mu(\{(x_1, \ldots, x_n) \in \Delta_p \ ; \ (\sum_{i=1}^n |x_i|^q)^{1/q} \le trr_q\}) dr$$

$$= n \int_0^1 r^{n-1} P_r dr$$

where

$$P_r = P\left((\sum_{i=1}^n |X_i|^q)^{1/q} (\sum_{i=1}^n |X_i|^p)^{-1/p} \le rt\frac{r_q}{r_p}\right).$$

Now, for all $0 < s < \infty$, $\mathbf{E}X_i^s = \Gamma\left(\frac{s+1}{p}\right)/\Gamma(1/p)$. It thus follows from the law of large numbers that, when $n \to \infty$,

$$N^{1/p-1/q}(\sum_{i=1}^n |X_i|^q)^{1/q}(\sum_{i=1}^n |X_i|^p)^{-1/p} \longrightarrow \Gamma\left(\frac{q+1}{p}\right)^{1/q} \Gamma(1/p)^{1/p-1/q}/\Gamma(1+1/p)^{1/p}$$

almost surely. By Stirling formula, for all $0 < s < \infty$,

$$r_s n^{-1/s} \to (e^{1/s} s^{1/s} 2\Gamma(1+1/s))^{-1}.$$

Combining all of that we get that if $rtA_{p,q} < 1$, $P_r \to 0$ as $n \to \infty$, and, if $rtA_{p,q} > 1$, $P_r \to 1$. The conclusion of the theorem, in the case $0 < p, q < \infty$, follows now easily from the equation

$$\mathrm{Vol}(B_p^n \cap tB_q^n) = n \int_0^1 r^{n-1} P_r dr \ .$$

The case $p = \infty$ is easier. Let $X_1, \ldots, X_n$ be independent random variables each uniformly distributed over the interval $[-1/2, 1/2]$, then

$$\mathrm{Vol}(B_\infty^n \cap tB_q^n) = P\left((\sum_{i=1}^n |X_i|^q)^{1/q} \le tr_q\right).$$

Now, $\mathbf{E}|X_i|^q = 2^{-q}(q+1)^{-1}$ so that by the strong law of large numbers

$$n^{-1/q}(\sum_{i=1}^n |X_i|^q)^{1/q} \to 2^{-1}(q+1)^{-1/q}.$$

Since as before $n^{-1/q} r_q \to (e^{1/q} q^{1/q} 2\Gamma(1+1/q))^{-1}$ we get that, as $n \to \infty$,

$$\mathrm{Vol}(B_\infty^n \cap tB_q^n) \to 0 \ \text{ if } \ 2^{-1}(q+1)^{-1/q} > t(e^{1/q} q^{1/q} 2\Gamma(1+1/q))^{-1}$$

and

$$\mathrm{Vol}(B_\infty^n \cap tB_q^n) \to 1 \ \text{ if } \ 2^{-1}(q+1)^{-1/q} < t(e^{1/q} q^{1/q} 2\Gamma(1+1/q))^{-1}. \qquad \blacksquare$$

Proof of Proposition. <u>a</u> Put, for $0 < y \le x < \infty$,

$$A(x,y) = e^{x-y} \frac{\Gamma(1+x)^{1+y}}{\Gamma(1+y)\Gamma(x+x/y)^y} \left(\frac{x}{y}\right)^y.$$

We need to prove that $A(x,y) < 1$ for $y < x$. Since clearly $A(y,y) = 1$, It is enough to show that, for $0 < y < x$,

$$\frac{\partial}{\partial x} \log A(x,y) < 0. \tag{3}$$

Putting $\phi(z) = \log \Gamma(z)$,

$$\frac{\partial}{\partial x} \log A(x,y) = 1 - \frac{y}{x} + (1+y)\phi'(1+x) - (1+y)\phi'(x + \frac{x}{y}).$$

Now,

$$\phi'(z) = \lim_{k \to \infty} \left(\log k - \sum_{i=1}^{k} \frac{1}{z+i-1} \right)$$

(see [A], p.17). So (3) translates to

$$(1+y) \sum_{i=1}^{\infty} \frac{1}{(x+i)(xy+x+iy-y)} - \frac{1}{x} > 0$$

or

$$\sum_{i=1}^{\infty} \frac{1}{(x+i)(y+1+\frac{iy}{x}-\frac{y}{x})} > \frac{1}{1+y}. \tag{4}$$

Now, the lefthand side in (4) is larger than

$$\int_0^{\infty} \frac{1}{(x+t+1)(y+1+ty/x)} dt = \frac{1}{1-y/x} \log\left(\frac{(1+y)}{y+y/x}\right)$$

$$= \frac{1}{1-y/x} (\log(y+1) - \log(y+y/x))$$

which, by the concavity of log and the fact that $y/x < 1$ is larger than

$$\frac{d}{dy} \log(1+y) = \frac{1}{1+y}.$$

<u>b</u> To prove that $A_{\infty,q} < 0$ we need to show that the function

$$A(x) = (1+x)^x e^{-x} \Gamma(1+x)^{-1}$$

is smaller than 1 on $(0,\infty)$. Now $\lim_{x \to 0} f(x) = 1$ and

$$\frac{d}{dx} \log f(x) = \log(1+x) - \lim_{k \to \infty} \left((\log k - \sum_{i=2}^{k} \frac{1}{x+i} \right)$$

which is easily seems to be negative on $(0,\infty)$. ∎

Remark. One can also prove in a similar fashion that $A_{p,q} < 1$ for $1 \leq q < p < \infty$.

Question. What is the asymptotic behaviour of $\mathrm{Vol}(B_p^n \cap tB_q^n)$ for $t = A_{p,q}^{-1}$?

References

[A] E. Artin, The Gamma Function, Holt, New York (1964).

[Mi] H. Minkowski, Geometrie der Zahlen, Chelsea, New York (1953).

[Mo] W. Moser, Research problems in discrete geometry. Sixth edition, Dept of Math. McGill Univ., Montreal (1981).

[SZ] G.Schechtman and J.Zinn, On the volume of the intersection of two L_p^n balls, Proc. AMS, to appear.

[S] C. L. Siegel, Lectures on the Geometry of Numbers, Springer-Verlag, Berlin (1989).

[W] B. Weißbach, Zu einer aufgabe von J. M. Wills, Acta Math. Hung. 48 (1986), 131-137.

ON THE RESTRICTION AND MULTIPLIER PROBLEMS IN R³

Jean Bourgain

IHES, France

and

University of Illinois, USA

1. Introduction

This paper is a follow-up of [B] and aims to sharpen some of the results obtained there, with special emphasis on the 3-dimensional case. Thus the presentation here is not self-contained and the reader should consult [B] for the proof so certain facts quoted below. We consider the following two problems (see again [B] for background material).

Problem 1. *Let μ be a measure carried by the 2-sphere S_2 such that $\left|\frac{d\mu}{d\sigma}\right| < C$, where σ is the surface measure. For which p is the Fourier transform $\hat{\mu}$ in $L^p(\mathbf{R}^3)$?*

If $\frac{d\mu}{d\sigma} \in L^2(S_2)$, one has $\hat{\mu} \in L^4(\mathbf{R}^3)$, which is an optimal result. In [B], it is shown that $\hat{\mu} \in L^p(\mathbf{R}^3)$ for $p > \frac{31}{8}$, provided $\frac{d\mu}{d\sigma} \in L^\infty(S_2)$. It is a conjecture that one may take in fact $p > 3$.

Problem 2. *Consider the Bochner-Riesz multiplier*

$$
\left.
\begin{aligned}
m_\lambda(\xi) &= \left(1 - |\xi|^2\right)^\lambda \quad \text{if } |\xi| \le 1 \\
&= \quad 0 \qquad\qquad \text{if } |\xi| > 1
\end{aligned}
\right\} .
\tag{1.1}
$$

For which $\lambda > 0$ is m_λ a bounded Fourier multiplier on $L^p(\mathbf{R}^3)$?

This problem is, of course, self-dual and the condition $\frac{3}{2+\lambda} < p < \frac{3}{1-\lambda}$ is known to be necessary. If we assume $p_* = p \vee p' \ge 4$, it is also sufficient. The conjecture implies in particular that m_λ if bounded for all $\lambda > 0$ in the range $p \in [\frac{3}{2}, 3]$. In [B], we proved that m_λ is bounded if

$$
\lambda > 1 - \frac{3}{p_*}
\tag{1.2}
$$

and $p_* \ge \frac{127}{32}$ (< 4). Our purpose here is to improve on this restriction, and get the result for $p_* \ge \frac{296}{75}$ (see the theorem at the end of the paper).

The problems mentioned above are intimately related to the behaviour of certain maximal functions, which in [B] were called the Kakeya maximal function f^*_δ and the Nikodym maximal function f^{**}_δ. We recall their definition

$$f^*_\delta(\xi) = \sup_T \frac{1}{|T|} \int_T |f| \tag{1.3}$$

defined for $\xi \in S_2$ and where the supremum is taken over all tubes T of unit length in direction ξ and of width $\delta > 0$. Here $|T|$ stands for the measure of T.

The Nikodym maximal function is defined by

$$f^{**}_\delta(x) = \sup_T \frac{1}{|T|} \int_T |f| \tag{1.4}$$

where $x \in \mathbf{R}^3$ and the supremum taken over all δ-tubes T centered at the point x. The conjectured estimates on $f^*_\delta, f^{**}_\delta$ are the following ($p \leq 3$)

$$\||f^*_\delta\|_{L^p(S_2)} \ll \left(\frac{1}{\delta}\right)^{\frac{3}{p}-1+\epsilon} \|f\|_{L^p(\mathbf{R}^3)} \tag{1.5}$$

$$\|f^{**}_\delta\|_{L^p(\mathbf{R}^3)} \ll \left(\frac{1}{\delta}\right)^{\frac{3}{p}-1+\epsilon} \|f\|_{L^p(\mathbf{R}^3)} . \tag{1.6}$$

Both inequalities (1.5), (1.6) were shown in [B] to hold for $p \leq \frac{7}{3}$. The validity of (1.5), (1.6) for $p = 3$ would imply in particular that so-called Kakeya and Nikodym sets in $\mathbf{R}^3$ are of maximal Hausdorff dimension (which is an open problem in any dimension > 2).

The logic of the approach here is the same as in [B] and may thus be summarized as follows:

$$(1.5) \implies \text{NON-}L^2 \text{ restriction theorems}$$
$$+$$
$$(1.6)$$

$$\implies \text{multiplier results}$$
$$\text{related to Problem 2.}$$

For later purposes, we consider the following variant of the Nikodym-maximal function, where the directions are restricted to an angular region $\Theta \subset S_2$

$$f^{**}_{f,\Theta}(x) = \sup_T \frac{1}{|T|} \int_T |f| \tag{1.7}$$

where the supremum is taken over all tubes T centered at x of unit length in direction $\xi \in \Theta$. We assume the size γ of the cap Θ to be $\geq \delta$. There is the following maximal inequality:

Lemma 1.8. *For $p \leq \frac{7}{3}$, Θ of size $\gamma > \delta$*

$$\|f_{\delta,\Theta}^{**}\|_p \ll \left(\frac{\gamma}{\delta}\right)^{\frac{3}{p}-1+\epsilon} \|f\|_p . \tag{1.9}$$

Proof: Let x_1, x_2, x_3 be $\mathbb{R}^3$-coordinates and assume Θ centered at the e_3-unit vector. The problem clearly reduces to functions supported by the cylinder

$$\{(x_1, x_2) \mid x_1^2 + x_2^2 < \gamma^2\} \times [|x_3| \leq 1] . \tag{1.10}$$

Consider the following coordinate change

$$\begin{cases} x_1 = \gamma x_1' \\ x_2 = \gamma x_2' \\ x_3 = x_3' . \end{cases} \tag{1.11}$$

The cylinder (1.10) is essentially carried to the unit cube and a δ-tube T centered at 0 in direction $\xi \in \Theta$ becomes a $\frac{\delta}{\gamma}$-tube in T'.

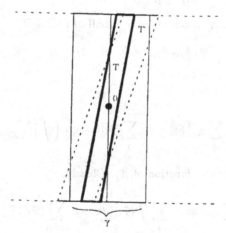

Put $g(x') = f(x)$, for which

$$\|g\|_p = \gamma^{-2/p}\|f\|_p . \tag{1.12}$$

If T' corresponds to the (δ, Θ)-tube T centered at 0, one has

$$\frac{1}{|T|}\int_T |f(x+y)|dy = \frac{1}{|T'|}\int_{T'} |g(x'+y')|dy' \leq g_{\delta/\gamma}^{**}(x') \tag{1.13}$$

and hence

$$f_{\delta,\Theta}^{**}(x) \leq g_{\delta/\gamma}^{**}(x') . \tag{1.14}$$

Therefore, by (1.6), (1.12),

$$\|f_{\delta,\Theta}^{**}\|_p \le \gamma^{2/p} \|g_{\delta/\gamma}^{**}\|_p \ll \gamma^{2/p} \left(\frac{\gamma}{\delta}\right)^{\frac{3}{p}-1+\varepsilon} \|g\|_p = \left(\frac{\gamma}{\delta}\right)^{\frac{3}{p}-1+\varepsilon} \|f\|_p \tag{1.15}$$

which is (1.9) . □

2. Fourier Transforms of Measures of Bounded Density Carried by the Sphere

By organizing the calculation of Section 6 of [B] a bit more efficiently we show

Proposition 2.1. *Let μ be a measure carried by S_2 with $\frac{d\mu}{d\sigma} \in L^\infty(S_2)$. Then $\widehat{\mu} \in L^p(\mathbf{R}^3)$ for* $p > \frac{58}{15}$.

Let $\varphi = \frac{d\mu}{d\sigma}$ about which we assume $|\varphi| \le 1$. Let $0 < \lambda < 1$ and consider the set $A = [\operatorname{Re}\widehat{\mu} > \lambda]$, which measure has to be estimated. Fix $R > 1$ to be specified later and consider subsets A_j of A satisfying

$$\sum |A_j| \sim |A| \tag{2.2}$$

$$A_j \text{ is contained in a ball of radius } R \tag{2.3}$$

$$\operatorname{dist}(A_{j_1}, A_{j_2}) > R \quad \text{if} \quad j_1 \ne j_2 . \tag{2.4}$$

Thus one has by (2.2)

$$\lambda|A| < c|\langle \widehat{\mu}, \sum_j \chi_j \rangle| \le c \int |\sum_j \widehat{\chi}_j| d\sigma < c \left(\int |\sum \widehat{\chi}_j|^2 d\sigma\right)^{1/2} \tag{2.5}$$

where χ_j stands for the indicator function of A_j. Clearly,

$$\int |\sum_j \widehat{\chi}_j|^2 d\sigma = \sum_j \int |\widehat{\chi}_j|^2 d\sigma + 2 \sum_{j_1 \ne j_2} \langle \widehat{\chi}_{j_1}, \widehat{\chi}_{j_2} \rangle_\sigma . \tag{2.6}$$

Because of (2.3), one has

$$\int |\widehat{\chi}_j|^2 d\sigma = \int |\widehat{\chi}_j|^2 d\sigma_{1/R} < CR \int |\widehat{\chi}_j|^2 d\xi = CR|A_j| , \tag{2.7}$$

denoting $\sigma_{1/R} = \sigma * \varphi_{1/R}$ where $\varphi \in \mathcal{D}$ satisfies $\widehat{\varphi} = 1$ on $B(0,2)$. The cross terms in (2.6) are estimated using the decay estimate

$$\widehat{\sigma}(\xi) = O(|\xi|^{-1}) . \tag{2.8}$$

Thus for $j_1 \neq j_2$

$$|\langle \hat{\chi}_{j_1}, \hat{\chi}_{j_2} \rangle| \leq \iint \chi_{j_1}(x)\chi_{j_2}(x-y)|\hat{\sigma}(x)|\,dx\,dy \leq C\rho(j,j_2)^{-1}|A_{j_1}||A_{j_2}| \tag{2.9}$$

denoting

$$\rho(j_1, j_2) = \text{dist}(A_{j_1}, A_{j_2}) . \tag{2.10}$$

Hence, by (2.9), the second term in (2.6) is easily seen to be bounded by

$$|A| \cdot \sup_{\rho > R} \rho^{-1+\epsilon} \sup_{z \in \mathbf{R}^3} \text{mes}\left[A \cap B(z,\rho)\right] . \tag{2.11}$$

At this point, we invoke Lemma 6.26 of [B] (actually using the stronger hypothesis $|\varphi| < 1$), yielding the estimate

$$\text{mes}\left[A \cap B(z,\rho)\right] \leq \lambda^{-q} \|\hat{\mu}\big|_{B(z,\rho)}\|_q^q \ll \lambda^{-q} \rho^{1-\frac{q}{4}+\epsilon} . \tag{2.12}$$

Here q is taken to satisfy $\left(\frac{q}{2}\right)' \leq p(3) = \frac{7}{3}$. Thus $q = \frac{7}{2}$.

Collecting estimates (2.5), (2.6), (2.7), (2.11), (2.12), it follows that

$$\lambda^2 |A|^2 \ll R|A| + \lambda^{-\frac{7}{2}} R^{-\frac{7}{8}+\epsilon} |A| . \tag{2.13}$$

An optimal choice of R yields thus

$$|A| < \lambda^{-\frac{58}{15}+\epsilon} , \tag{2.14}$$

which proves Proposition 2.1. □

Since the map $L^\infty(S_2) \to L^p(\mathbf{R}^3) : \varphi \mapsto \widehat{\varphi\sigma}$ is bounded for $p > \frac{58}{15}$, we get from general factorization theory (see [P]), together with the invariance under orthogonal transformation:

Proposition 2.15. *Let* $p > \frac{58}{15}$ *and* μ *a measure carried by* S_2 *such that* $\frac{d\mu}{d\sigma} \in L^p(S_2)$. *Then* $\hat{\mu} \in L^p(\mathbf{R}^2)$.

Remark. The previous considerations apply equally well in higher dimensions.

3. An L^4-estimate for S_2-carried Densities

It follows from the L^2-restriction theorem that if $\mu \in M(S_2)$ and $\varphi = \frac{d\mu}{d\sigma} \in L^2(S_2)$, then $\hat{\mu} \in L^4(\mathbf{R}^3)$ and

$$\|\hat{\mu}\|_4 \leq C\left[\int |\varphi|^2 d\sigma\right]^{1/2} . \tag{3.1}$$

In this section we give a different estimate of $\|\hat{\mu}\|_4$, which in some sense is more precise. It will give a better result in an interpolation argument appearing in the study of Bochner-Riesz multipliers.

Let $\Omega \subset S_2$ be a measurable subset of the sphere and take for $\varphi = \chi_\Omega$. Fix $R > 1$. Clearly

$$\|\hat{\mu}\|_{L^4(B(0,R))}^4 \leq CR^3 \int_{\Omega^4} \rho(x_1 + x_2 - x_3 - x_4)\sigma(dx_1)\sigma(dx_2)\sigma(dx_3)\sigma(dx_4) \qquad (3.2)$$

where ρ stands for the indicator function of the ball $B(0, \frac{1}{R})$. If $\rho(x_1 + x_2 - x_3 - x_4) \neq 0$, one has for $i = 3$ or $i = 4$ the following inequalities:

$$\left| |x_1 + x_2 - x_i| - 1 \right| < \frac{1}{R}, \qquad (3.3)$$

$$\left| 1 + \langle x_1, x_2 \rangle - \langle x_1, x_i \rangle - \langle x_2, x_i \rangle \right| < \frac{C}{R}, \qquad (3.4)$$

$$\left| |x_i - \frac{x_1 + x_2}{2}|^2 - |\frac{x_1 - x_2}{2}|^2 \right| < \frac{C}{R}, \qquad (3.5)$$

$$\left| |x_i - \frac{x_1 + x_2}{2}| - |\frac{x_1 - x_2}{2}| \right| < \frac{C}{R|x_1 - x_2|}. \qquad (3.6)$$

Also

$$|x_1 - x_3| + |x_1 - x_4| > |x_1 - x_2| - \frac{1}{R}. \qquad (3.7)$$

Restricting for given $x_1 \in \Omega$ to $|x_1 - x_2| < \frac{1}{R^{1/2}}$ yields by (3.5)

$$\left| x_3 - \frac{x_1 + x_2}{2} \right| < \frac{C}{R^{1/2}} \qquad (3.8)$$

and the contribution to the right member of (3.2) is bounded by

$$C \cdot R^3 \cdot |\Omega| \cdot \frac{1}{R} \cdot \frac{1}{R} \cdot \frac{1}{R^2} < C \cdot \frac{|\Omega|}{R}. \qquad (3.9)$$

If $|x_1 - x_2| > \frac{1}{R^{1/2}}$, either $|x_1 - x_3| > \frac{|x_1 - x_2|}{2}$ or $|x_1 - x_4| > \frac{|x_1 - x_2|}{2}$, by (3.7). We consider the contribution of

$$\left\{ (x_1, x_2, x_3, x_4) \in \Omega^4 \mid \rho(x_1 + x_2 - x_3 - x_4) = 1, \quad |x_1 - x_3| > \frac{|x_1 - x_2|}{2} \right\} \qquad (3.10)$$

which measure may be bounded by

$$R^{-2} \int_\Omega \mathrm{mes}_{\Omega \times \Omega} \left\{ (x_2, x_3) \mid \left| |x_3 - \frac{x_1 + x_2}{2}| - |\frac{x_1 - x_2}{2}| \right| < \frac{C}{R|x_1 - x_2|} \right.$$

$$\left. \text{and} \quad |x_1 - x_3| \gtrsim |x_1 - x_2| > \frac{1}{R^{1/2}} \right\} \qquad (3.11)$$

where the integration is performed with respect to x_1.

Let δ range dyadically over the interval $[R^{-1/2}, 1]$. For each δ, consider a covering $\mathcal{C}_\delta$ of S_2 by caps of size δ, $\mathcal{C}_\delta$ having bounded multiplicity. Estimate (3.11) by

$$R^{-2} \sum_{\delta \text{ dyadic}} \sum_{\tau \in \mathcal{C}_\delta} |\Omega \cap \tau| \sup_{x \in \tau} \text{mes} \left\{ (x_2, x_3) \in \Omega' \times \Omega' \mid \right.$$

$$\left. \left| |x_3 - \frac{x + x_2}{2}| - |\frac{x - x_2}{2}| \right| \leq \frac{1}{R\delta} \; ; \; |x - x_2| \sim \delta \sim |x - x_3| \right\} \tag{3.12}$$

denoting

$$\Omega' = \Omega \cap \tau_\delta \tag{3.13}$$

$$\tau_\delta = \delta - \text{neighborhood of } \tau \tag{3.14}$$

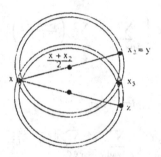

Define

$$\Gamma = \left\{ (x_2, x_3) \in S_2 \times X_2 \mid \left| |x_3 - \frac{x + x_2}{2}| - |\frac{x - x_2}{2}| \right| < \frac{1}{R\delta} \; ; \; |x - x_2| \sim \delta \sim |x - x_3| \right\} \tag{3.15}$$

and estimate

$$|(\Omega' \times \Omega') \cap \Gamma| \leq |\Omega'|^{1/2} \left\{ \int_{\Omega'} \left[\int_{\Omega'} \chi_\Gamma(x_2, x_3) dx_2 \right]^2 dx_3 \right\}^{1/2} \tag{3.16}$$

$$= |\Omega'|^{1/2} \left\{ \iint_{\Omega' \times \Omega'} \text{mes} \left[\Gamma(y, \cdot) \cap \Gamma(z, \cdot) \right] dy \, dz \right\}^{1/2} . \tag{3.17}$$

The set $\Gamma(y, \cdot) \cap \Gamma(z, \cdot)$ is the intersection of two annuli on S^2. The size of each annulus is of order δ, the width $\frac{1}{R\delta} \ll \delta$. They intersect at a given point x and at some point x_3, where

$|x - x_3| \sim \delta$. (See above picture.) An easy calculation yields therefore that

$$\text{mes}\left[\Gamma(y, \cdot) \cap \Gamma(z, \cdot)\right] \sim \left(\frac{1}{R\delta}\right)^2 \left|\text{angle}\left([x, y], [x, z]\right)\right|^{-1} \tag{3.18}$$

$$\sim \left(\frac{1}{R\delta}\right)^2 \frac{\delta}{|y - z|} . \tag{3.19}$$

Substituting (3.19) in (3.17) yields

$$|\Omega'|^{1/2} \frac{1}{R\delta^{1/2}} \left\{ \iint\limits_{\Omega' \times \Omega'} |y - z|^{-1} \sigma(dy) \sigma(dz) \right\}^{1/2} \leq \frac{1}{R\delta^{1/2}} |\Omega'|^{5/4} . \tag{3.20}$$

By (3.12) and (3.9), one obtains the following bound for (3.2)

$$\sum_{\substack{\delta \text{ dyadic} \\ 1 > \delta > R^{-1/2}}} \delta^{-1/2} \sum_{\tau \in \mathcal{C}_\delta} |\Omega \cap \tau_\delta|^{9/4} + \frac{|\Omega|}{R} . \tag{3.21}$$

Consequently, letting $R \to \infty$, one finds that

$$\|\widehat{\mu}\|_4 \leq C \left\{ \sum_{\delta \text{ dyadic}} \delta^4 \sum_{\tau \in \mathcal{C}_\delta} \left(\frac{|\Omega \cap \tau|}{|\tau|}\right)^{9/4} \right\}^{1/4} \tag{3.22}$$

where $\frac{d\mu}{d\sigma} = \chi_\Omega$. Thus

Lemma 3.23. *Let* $\Omega \subset S_2$, $d\mu = \chi_\Omega d\sigma$. *Then*

$$\|\widehat{\mu}\|_4 \leq C \left\{ \sum_{\delta \text{ dyadic}} \delta^4 \sum_{\tau \in \mathcal{C}_\delta} \|\chi_\Omega|_\tau\|_{L^{16/9}(\tau)}^4 \right\} \tag{3.23}$$

where $L^p(\tau)$ *refers to the normalized measure on the cap* τ.

Our next purpose is to interpolate between (2.15) and (3.23). In this respect, the norm defined by (3.23) leads to a problem, since one considers only a subspaces of the $\ell_{L^{16/9}}^4$-space associated to the (3.23)-norm. In fact, we will only perform this interpolation in a "crude" way, which will suffice for later purposes.

Assume $\Omega \subset S_2$ is a union of ε-caps. If $d\mu = \chi_\Omega d\sigma$, (3.23) clearly yields

$$\|\widehat{\mu}\|_4 \leq C \left\{ \sum_{\substack{\delta \text{ dyadic} \\ \delta > \varepsilon}} \delta^4 \sum_{\tau \in \mathcal{C}_\delta} \|\chi_\Omega|_\tau\|_{L^{16/9}(\tau)}^4 \right\}^{1/4} . \tag{3.24}$$

One may then find some $\varepsilon < \delta' < 1$ and a subset $\mathcal{C}'$ of $\mathcal{C}_{\delta'}$ such that if we denote

$$\Omega' = \Omega \cap \bigcup_{\tau \in \mathcal{C}'} \tau , \tag{3.25}$$

the following properties hold

$$(3.24) < \left(\log\frac{1}{\varepsilon}\right)\delta'\left(\sum_{\tau \in \mathcal{C}'}\|\chi_{\Omega'}|_\tau\|_{L^{16/9}(\tau)}^4\right)^{1/4}, \tag{3.26}$$

$$|\Omega' \cap \tau_1| \sim |\Omega' \cap \tau_2| \quad \text{for} \quad \tau_1, \tau_2 \in \mathcal{C}'. \tag{3.27}$$

This is straightforward and left to the reader.

Define

$$\frac{d\mu'}{d\sigma} = \chi_{\Omega'}. \tag{3.28}$$

Fix $p_0 > \frac{58}{15}$ as in (2.15) and $0 < \theta < 1$. Define p and r by

$$\frac{1}{p} = \frac{\theta}{p_0} + \frac{1-\theta}{4} \quad \text{and} \quad \frac{1}{r} = \frac{\theta}{p_0} + \frac{1-\theta}{16/9}. \tag{3.29}$$

Estimate by Hölder's inequality

$$\|\widehat{\mu}'\|_p \le \|\widehat{\mu}'\|_{p_0}^\theta \|\widehat{\mu}'\|_4^{1-\theta}. \tag{3.30}$$

By (2.15), (3.27), fixing some $\tau \in \mathcal{C}'$

$$\|\widehat{\mu}'\|_{p_0} \le C|\Omega'|^{1/p_0} \le C|\mathcal{C}'|^{1/p_0}|\Omega' \cap \tau|^{1/p_0}. \tag{3.31}$$

By (3.24) applied to Ω', (3.26), (3.27)

$$\begin{aligned}
\|\widehat{\mu}'\|_4 &\le C\left\{\sum_{\delta > \varepsilon}\delta^4\sum_{\tau \in \mathcal{C}_\delta}\|\chi_{\Omega'}|_\tau\|_{L^{16/9}(\tau)}^4\right\}^{1/4} \\
&\le C\left\{\sum_{\delta > \varepsilon}\delta^4\sum_{\tau \in \mathcal{C}_\delta}\|\chi_\Omega|_\tau\|_{L^{16/9}(\tau)}^4\right\}^{1/4} \\
&\le C\left(\left(\log\frac{1}{\varepsilon}\right)\cdot(\delta')^{-1/8}\cdot|\mathcal{C}'|^{1/4}|\Omega' \cap \tau|^{9/16}\right).
\end{aligned} \tag{3.32}$$

Thus, from (3.30), (3.31), (3,32), (3.29)

$$\|\widehat{\mu}'\|_p \le C\left(\log\frac{1}{\varepsilon}\right)\cdot(\delta')^{-1/8(1-\theta)}\cdot|\mathcal{C}'|^{1/p}|\Omega' \cap \tau|^{1/r} \tag{3.33}$$

$$\le C\left(\log\frac{1}{\varepsilon}\right)\left\{\sum_{\delta > \varepsilon}\delta^{-p/8(1-\theta)}\sum_{\tau \in \mathcal{C}_\delta}|\Omega \cap \tau|^{p/r}\right\}^{1/p}. \tag{3.34}$$

Replace next Ω by

$$\Omega_1 = \Omega \backslash \Omega' \tag{3.35}$$

and repeat previous considerations. Observe that by (3.26)

$$\left\{ \sum_\delta \delta^4 \sum_{\tau \in C_\delta} \|\chi_{\Omega_1}|_\tau\|^4_{L^{16/9}(\tau)} \right\}^{1/4}$$

$$< \left[1 - \left(\log \frac{1}{\varepsilon} \right)^{-4} \right] \left\{ \sum_\delta \delta^4 \sum_\tau \|\chi_\Omega|_\tau\|^4_{L^{16/9}(\tau)} \right\}^{1/4} . \tag{3.36}$$

Thus the process may be stopped after at most $\sim \left(\log \frac{1}{\varepsilon} \right)^5$ steps. Summing up the consecutive contributions, each evaluated by (3.36), yields

$$\|\hat{\mu}\|_p \le \sum_\alpha \|\widehat{\mu_\alpha}\|_p < C \left(\log \frac{1}{\varepsilon} \right)^6 \left\{ \sum_{\delta > \varepsilon} \delta^{-p/8(1-\theta)} \sum_{\tau \in C_\delta} |\Omega \cap \tau|^{p/\tau} \right\}^{1/p} . \tag{3.37}$$

As in [B], we will use a discrete version of (3.37).

Lemma 3.38. Let $p_0 = \frac{58}{15}$, $0 < \theta < 1$ and define p and r by

$$\frac{1}{p} = \frac{\theta}{p_0} + \frac{1-\theta}{4} , \qquad \frac{1}{r} = \frac{\theta}{p_0} + \frac{1-\theta}{16/9} . \tag{3.39}$$

Let $R > 1$ and $\{x_\alpha\}$ a $\frac{1}{R}$-separated system of points on S_2. Then

$$\left\| \sum_\alpha a_\alpha e^{2\pi i \langle \xi, x_\alpha \rangle} \right\|_{L^p(B(0,R))}$$

$$\ll R^{2/r'+\varepsilon} \left\{ \sum_{\delta > \frac{1}{R}} \delta^{-p/8(1-\theta)} \sum_{\tau \in C_\delta} \|\{a_\alpha \mid x_\alpha \in \tau\}\|^p_r \right\}^{1/p} . \tag{3.40}$$

Proof: It clearly suffices to consider the case of $0, 1$-valued coefficients. Since ξ is restricted to $B(0, R)$, the left member of (3.40) may roughly speaking be replaced by

$$R^2 \int_\Omega e^{2\pi i \langle \xi, x \rangle} \sigma(dx) \tag{3.41}$$

where Ω stands for a $\frac{1}{R}$-neighborhood of $\{x_\alpha \mid a_\alpha = 1\}$. It remains to invoke (3.37). □

4. Bochner-Riesz Summability

We consider the multiplier (1.1)

$$m_\lambda(\xi) = (1 - |\xi|^2)^\lambda \quad \text{if} \quad |\xi| \le 1$$
$$= 0 \qquad \text{otherwise} \tag{4.1}$$

acting as Fourier multiplier on $L^p(\mathbf{R}^3)$.

Decomposing the ball $B(0, 1)$ in shells $1 - \delta < |\xi| < 1 - \frac{\delta}{2}$, the multiplier m_λ is constructed by considering $(\sqrt{\delta} \times \sqrt{\delta} \times \delta)$-boxes B_α as indicated below (corresponding to the 2-dimensional picture).

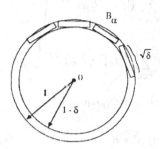

One may build a multiplier with spherical symmetry from $\sum_\alpha \chi_{B_\alpha}$ by averaging over the orthogonal group. The value of λ that will make (4.1) bounded corresponds to the condition

$$\left\| \sum (\hat{f} \cdot B_\alpha)^\vee \right\|_p < C \cdot \delta^{-\lambda + \varepsilon} \|f\|_p . \tag{4.2}$$

The left member of (4.2) is studied as in section 7 of [B], but using the new inequality (3.40). Thus we first partition $\mathbf{R}^3$ in boxes Q_j of size $R \sim \delta^{-1/2}$ on each of which $\sum (\hat{f} \cdot B_\alpha)^\vee$ behaves like $\sum_\alpha a_\alpha e^{2\pi i \langle x_\alpha, \xi \rangle}$, where x_α denotes the center of B_α and $a_\alpha \sim \int_{B_\alpha} \hat{f}(x) e^{2\pi i \langle x - x_\alpha, \xi \rangle} \sigma(dx)$, $\xi \in Q_j$. Thus (3.40) yields

$$\int_{Q_j} \left| \sum (\hat{f} \cdot B_\alpha)^\vee \right|^0 \ll$$

$$\delta^{-\frac{p}{r} - \varepsilon} \sum_{\gamma > \delta^{1/2}} \gamma^{-\frac{p}{8}(1-\theta)} \sum_{\tau \in C_\gamma} \left(\sum_{x_\alpha \in \tau} (\hat{f} \cdot B_\alpha)^\vee (\xi)|^r \right)^{p/r} \tag{4.3}$$

$$\sim \delta^{-\frac{p}{r} + \frac{3}{2} - \varepsilon} \sum_{\gamma > \delta^{1/2}} \gamma^{-\frac{p}{8}(1-\theta)} \sum_{\tau \in C_\gamma} \int_{Q_j} \left(\sum_{x_\alpha \in \tau} |(\hat{f} \cdot B_\alpha)^\vee|^r \right)^{p/r} \tag{4.4}$$

and summation over the Q_j gives

$$\int \left| \sum (\hat{f} \cdot B_\alpha)^\vee \right|_p \ll$$

$$\delta^{-\frac{p}{r} + \frac{3}{2} - \varepsilon} \sum_{\gamma > \delta^{1/2}} \gamma^{-\frac{p}{8}(1-\theta)} \sum_{\tau \in C_\gamma} \int \left(\sum_{x_\alpha \in \tau} |(\hat{f} \cdot B_\alpha)^\vee|^r \right)^{p/r} . \tag{4.5}$$

As in [B], with $(\hat{f} \cdot B_\alpha)^\vee = f_\alpha * (\chi_{B_\alpha})^\vee$ where $\hat{f}_\alpha = \hat{f}\big|_{Q_\alpha}$ and Q_α is the $\sqrt{\delta}$-size cube centered at x_α. Estimate further

$$\left| f_\alpha * (\chi_{B_\alpha})^\vee \right| \leq |f_\alpha| * K_\alpha \tag{4.6}$$

where K_α is a tube centered at 0, has direction x_α and excentricity $\delta^{1/2}$ (the K_α's have the shape of $(\chi_{B_\alpha})^\vee$ and may be chosen to reconstruct $(\chi_{B_\alpha})^\vee$). Proceeding as in [B], one estimates for given $\tau \in \mathcal{C}_\gamma$

$$\left[\int \left[\sum_{x_\alpha \in \tau} (|f_\alpha| * K_\alpha)^r \right]^{p/r} \right]^{r/p}$$

$$\leq \left[\int \left[\sum_{x_\alpha \in \tau} (|f_\alpha|^r * K_\alpha) \right]^{p/r} \right]^{r/p} \tag{4.7}$$

by duality, hence for some $g \in L^{(p/r)'}(\mathbf{R}^3)$, $\|g\|_{(p/r)'} = 1$

$$(4.7) = \sum_{x_\alpha \in \tau} \langle |f_\alpha|^r * K_\alpha, g \rangle$$

$$\leq \left\| \sum_{x_\alpha \in \tau} |f_\alpha|^r \right\|_{p/r} \left\| \max_{x_\alpha \in \tau}(g * K_\alpha) \right\|_{(p/r)'} . \tag{4.8}$$

We assume that r, p satisfy

$$r \geq 2, \qquad \left(\frac{p}{r}\right)' \leq \frac{7}{3} . \tag{4.9}$$

Then

$$\left\| \sum_{x_\alpha \in \tau} |f_\alpha|^r \right\|_{p/r} \leq \left\| \left(\sum_{x_\alpha \in \tau} |f_\alpha|^2 \right)^{1/2} \right\|_p^r . \tag{4.10}$$

After suitable rescaling, $\max_{x_\alpha \in \tau}(g * K_\alpha)$ leads to a Nikodym-maximal function involving tubes of width $\delta^{1/2}$ directed towards τ, which is of size γ. Hence, Lemma 1.8 applies in estimating the second factor of (4.8)

$$\left\| \max_{x_\alpha \in \tau}(g * K_\alpha) \right\|_{(p/r)'} \ll \left(\frac{\gamma}{\sqrt{\delta}} \right)^{\frac{3}{(\frac{p}{r})'} - 1 + \varepsilon} . \tag{4.11}$$

From (4.7), (4.8), (4.10), (4.11), the following estimate on (4.5) is derived

$$\delta^{-\frac{p}{r} + \frac{3}{2} - \frac{1}{2}\frac{p}{r}\left[\frac{3(p-r)}{p} - 1\right] - \varepsilon} .$$

$$\sum_{\substack{1 > \gamma > \delta^{1/2} \\ \gamma \text{ dyadic}}} \gamma^{-\frac{p}{8}(1-\theta) + \frac{p}{r}\left[\frac{3(p-r)}{p} - 1\right]} \sum_{\tau \in \mathcal{C}_\gamma} \left\| \left(\sum_{x_\alpha \in \tau} |f_\alpha|^2 \right)^{1/2} \right\|_p^p . \tag{4.12}$$

Recall (3.39)

$$\frac{1}{p} = \frac{\theta}{58/15} + \frac{1-\theta}{4} , \qquad \frac{1}{r} = \frac{\theta}{58/15} + \frac{1-\theta}{16/9} . \tag{4.13}$$

Take

$$\theta = \frac{29}{74} \Rightarrow p = \frac{296}{75} , \qquad r = \frac{4}{7} \cdot p = \frac{1184}{525} > 2 \tag{4.14}$$

satisfying (4.9). The γ exponent in (4.12) is given by

$$-\frac{296 \cdot 45}{75 \cdot 8 \cdot 74} + 2 \cdot \frac{7}{4} - 3 = \frac{1}{5} > 0 . \tag{4.15}$$

Consequently (4.12) is bounded by

$$\delta^{3-p-\varepsilon} \left(\sum_\gamma \gamma^{1/5} \right) \int \left(\sum_\alpha |f_\alpha|^2 \right)^{p/2} < C \cdot \delta^{3-p-\varepsilon} \|f\|_p^p . \tag{4.16}$$

Thus the number λ in (4.2) may be taken $1 - \frac{3}{p}$, which is the Bochner-Riesz exponent. Hence, we have

Theorem. m_λ *is a bounded Fourier multiplier on* $L^p(\mathbb{R}^3)$ *provided*

$$p \in \left(\frac{3}{2+\lambda}, \frac{3}{1-\lambda} \right) \setminus \left(\frac{296}{221}, \frac{296}{75} \right) .$$

References

[B]   J. Bourgain, Besicovitch type maximal functions and applications to Fourier Analysis, to appear in Geometric and Functional Analysis 1 (1991), N.2.

[P]   G. Pisier, Factorization of linear operators through $L^{p,1}$, $L^{p,\infty}$, Math. Nachrichten.

Lecture Notes in Mathematics

For information about Vols. 1–1272
please contact your bookseller or Springer-Verlag

Vol. 1318: Y. Felix (Ed.), Algebraic Topology – Rational Homotopy. Proceedings, 1986. VIII, 245 pages. 1988.

Vol. 1319: M. Vuorinen, Conformal Geometry and Quasiregular Mappings. XIX, 209 pages. 1988.

Vol. 1320: H. Jürgensen, G. Lallement, H.J. Weinert (Eds.), Semigroups, Theory and Applications. Proceedings, 1986. X, 416 pages. 1988.

Vol. 1321: J. Azéma, P.A. Meyer, M. Yor (Eds.), Séminaire de Probabilités XXII. Proceedings. IV, 600 pages. 1988.

Vol. 1322: M. Métivier, S. Watanabe (Eds.), Stochastic Analysis. Proceedings, 1987. VII, 197 pages. 1988.

Vol. 1323: D.R. Anderson, H.J. Munkholm, Boundedly Controlled Topology. XII, 309 pages. 1988.

Vol. 1324: F. Cardoso, D.G. de Figueiredo, R. Iório, O. Lopes (Eds.), Partial Differential Equations. Proceedings, 1986. VIII, 433 pages. 1988.

Vol. 1325: A. Truman, I.M. Davies (Eds.), Stochastic Mechanics and Stochastic Processes. Proceedings, 1986. V, 220 pages. 1988.

Vol. 1326: P.S. Landweber (Ed.), Elliptic Curves and Modular Forms in Algebraic Topology. Proceedings, 1986. V, 224 pages. 1988.

Vol. 1327: W. Bruns, U. Vetter, Determinantal Rings. VII, 236 pages. 1988.

Vol. 1328: J.L. Bueso, P. Jara, B. Torrecillas (Eds.), Ring Theory. Proceedings, 1986. IX, 331 pages. 1988.

Vol. 1329: M. Alfaro, J.S. Dehesa, F.J. Marcellan, J.L. Rubio de Francia, J. Vinuesa (Eds.): Orthogonal Polynomials and their Applications. Proceedings, 1986. XV, 334 pages. 1988.

Vol. 1330: A. Ambrosetti, F. Gori, R. Lucchetti (Eds.), Mathematical Economics. Montecatini Terme 1986. Seminar. VII, 137 pages. 1988.

Vol. 1331: R. Bamón, R. Labarca, J. Palis Jr. (Eds.), Dynamical Systems, Valparaiso 1986. Proceedings. VI, 250 pages. 1988.

Vol. 1332: E. Odell, H. Rosenthal (Eds.), Functional Analysis. Proceedings. 1986–87. V, 202 pages. 1988.

Vol. 1333: A.S. Kechris, D.A. Martin, J.R. Steel (Eds.), Cabal Seminar 81–85. Proceedings, 1981–85. V, 224 pages. 1988.

Vol. 1334: Yu.G. Borisovich, Yu.E. Gliklikh (Eds.), Global Analysis – Studies and Applications III. V, 331 pages. 1988.

Vol. 1335: F. Guillén, V. Navarro Aznar, P. Pascual-Gainza, F. Puerta, Hyperrésolutions cubiques et descente cohomologique. XII, 192 pages. 1988.

Vol. 1336: B. Helffer, Semi-Classical Analysis for the Schrödinger Operator and Applications. V, 107 pages. 1988.

Vol. 1337: E. Sernesi (Ed.), Theory of Moduli. Seminar, 1985. VIII, 232 pages. 1988.

Vol. 1338: A.B. Mingarelli, S.G. Halvorsen. Non-Oscillation Domains of Differential Equations with Two Parameters. XI, 109 pages. 1988.

Vol. 1339: T. Sunada (Ed.), Geometry and Analysis of Manifolds. Proceedings, 1987. IX, 277 pages. 1988.

Vol. 1340: S. Hildebrandt, D.S. Kinderlehrer, M. Miranda (Eds.), Calculus of Variations and Partial Differential Equations. Proceedings, 1986. IX, 301 pages. 1988.

Vol. 1341: M. Dauge, Elliptic Boundary Value Problems on Corner Domains. VIII, 259 pages. 1988.

Vol. 1342: J.C. Alexander (Ed.), Dynamical Systems. Proceedings, 1986–87. VIII, 726 pages. 1988.

Vol. 1343: H. Ulrich, Fixed Point Theory of Parametrized Equivariant Maps. VII, 147 pages. 1988.

Vol. 1344: J. Král, J. Lukes, J. Netuka, J. Vesely´ (Eds.), Potential Theory – Surveys and Problems. Proceedings, 1987. VIII, 271 pages. 1988.

Vol. 1345: X. Gomez-Mont, J. Seade, A. Verjovski (Eds.), Holomorphic Dynamics. Proceedings, 1986. VII. 321 pages. 1988.

Vol. 1346: O.Ya. Viro (Ed.), Topology and Geometry – Rohlin Seminar. XI, 581 pages. 1988.

Vol. 1347: C. Preston, Iterates of Piecewise Monotone Mappings on an Interval. V, 166 pages. 1988.

Vol. 1348: F. Borceux (Ed.), Categorical Algebra and its Applications. Proceedings, 1987. VIII, 375 pages. 1988.

Vol. 1349: E. Novak, Deterministic and Stochastic Error Bounds in Numerical Analysis. V, 113 pages. 1988.

Vol. 1350: U. Koschorke (Ed.), Differential Topology Proceedings, 1987, VI, 269 pages. 1988.

Vol. 1351: I. Laine, S. Rickman, T. Sorvali (Eds.), Complex Analysis, Joensuu 1987. Proceedings. XV, 378 pages. 1988.

Vol. 1352: L.L. Avramov, K.B. Tchakerian (Eds.), Algebra – Some Current Trends. Proceedings. 1986. IX, 240 Seiten. 1988.

Vol. 1353: R.S. Palais, Ch.-l. Teng, Critical Point Theory and Submanifold Geometry. X, 272 pages. 1988.

Vol. 1354: A. Gómez, F. Guerra, M.A. Jiménez, G. López (Eds.), Approximation and Optimization. Proceedings, 1987. VI, 280 pages. 1988.

Vol. 1355: J. Bokowski, B. Sturmfels, Computational Synthetic Geometry. V, 168 pages. 1989.

Vol. 1356: H. Volkmer, Multiparameter Eigenvalue Problems and Expansion Theorems. VI, 157 pages. 1988.

Vol. 1357: S. Hildebrandt, R. Leis (Eds.), Partial Differential Equations and Calculus of Variations. VI, 423 pages. 1988.

Vol. 1358: D. Mumford, The Red Book of Varieties and Schemes. V, 309 pages. 1988.

Vol. 1359: P. Eymard, J.-P. Pier (Eds.) Harmonic Analysis. Proceedings, 1987. VIII, 287 pages. 1988.

Vol. 1360: G. Anderson, C. Greengard (Eds.), Vortex Methods. Proceedings, 1987. V, 141 pages. 1988.

Vol. 1361: T. tom Dieck (Ed.), Algebraic Topology and Transformation Groups. Proceedings. 1987. VI, 298 pages. 1988.

Vol. 1362: P. Diaconis, D. Elworthy, H. Föllmer, E. Nelson, G.C. Papanicolaou, S.R.S. Varadhan. École d´ Été de Probabilités de Saint-Flour XV–XVII. 1985–87 Editor: P.L. Hennequin. V, 459 pages. 1988.

Vol. 1363: P.G. Casazza, T.J. Shura, Tsirelson´s Space. VIII, 204 pages. 1988.

Vol. 1364: R.R. Phelps, Convex Functions, Monotone Operators and Differentiability. IX, 115 pages. 1989.

Vol. 1365: M. Giaquinta (Ed.), Topics in Calculus of Variations. Seminar, 1987. X, 196 pages. 1989.

Vol. 1366: N. Levitt, Grassmannians and Gauss Maps in PL-Topology. V, 203 pages. 1989.

Vol. 1367: M. Knebusch, Weakly Semialgebraic Spaces. XX, 376 pages. 1989.

Vol. 1368: R. Hübl, Traces of Differential Forms and Hochschild Homology. III, 111 pages. 1989.

Vol. 1369: B. Jiang, Ch.-K. Peng, Z. Hou (Eds.), Differential Geometry and Topology. Proceedings, 1986–87. VI, 366 pages. 1989.

Vol. 1370: G. Carlsson, R.L. Cohen, H.R. Miller, D.C. Ravenel (Eds.), Algebraic Topology. Proceedings, 1986. IX, 456 pages. 1989.

Vol. 1371: S. Glaz, Commutative Coherent Rings. XI, 347 pages. 1989.

Vol. 1372: J. Azéma, P.A. Meyer, M. Yor (Eds.), Séminaire de Probabilités XXIII. Proceedings. IV, 583 pages. 1989.

Vol. 1373: G. Benkart, J.M. Osborn (Eds.), Lie Algebras. Madison 1987. Proceedings. V, 145 pages. 1989.

Vol. 1374: R.C. Kirby, The Topology of 4-Manifolds. VI, 108 pages. 1989.

Vol. 1375: K. Kawakubo (Ed.), Transformation Groups. Proceedings, 1987. VIII, 394 pages, 1989.

Vol. 1376: J. Lindenstrauss, V.D. Milman (Eds.), Geometric Aspects of Functional Analysis. Seminar (GAFA) 1987–88. VII, 288 pages. 1989.

Vol. 1377: J.F. Pierce, Singularity Theory, Rod Theory, and Symmetry-Breaking Loads. IV, 177 pages. 1989.

Vol. 1378: R.S. Rumely, Capacity Theory on Algebraic Curves. III, 437 pages. 1989.

Vol. 1379: H. Heyer (Ed.), Probability Measures on Groups IX. Proceedings, 1988. VIII, 437 pages. 1989.

Vol. 1380: H.P. Schlickewei, E. Wirsing (Eds.), Number Theory, Ulm 1987. Proceedings. V, 266 pages. 1989.

Vol. 1381: J.-O. Strömberg, A. Torchinsky, Weighted Hardy Spaces. V, 193 pages. 1989.

Vol. 1382: H. Reiter, Metaplectic Groups and Segal Algebras. XI, 128 pages. 1989.

Vol. 1383: D.V. Chudnovsky, G.V. Chudnovsky, H. Cohn, M.B. Nathanson (Eds.), Number Theory, New York 1985–88. Seminar. V, 256 pages. 1989.

Vol. 1384: J. Garcia-Cuerva (Ed.), Harmonic Analysis and Partial Differential Equations. Proceedings, 1987. VII, 213 pages. 1989.

Vol. 1385: A.M. Anile, Y. Choquet-Bruhat (Eds.), Relativistic Fluid Dynamics. Seminar, 1987. V, 308 pages. 1989.

Vol. 1386: A. Bellen, C.W. Gear, E. Russo (Eds.), Numerical Methods for Ordinary Differential Equations. Proceedings, 1987. VII, 136 pages. 1989.

Vol. 1387: M. Petkovi´c, Iterative Methods for Simultaneous Inclusion of Polynomial Zeros. X, 263 pages. 1989.

Vol. 1388: J. Shinoda, T.A. Slaman, T. Tugué (Eds.), Mathematical Logic and Applications. Proceedings, 1987. V, 223 pages. 1989.

Vol. 1000: Second Edition. H. Hopf, Differential Geometry in the Large. VII, 184 pages. 1989.

Vol. 1389: E. Ballico, C. Ciliberto (Eds.), Algebraic Curves and Projective Geometry. Proceedings, 1988. V, 288 pages. 1989.

Vol. 1390: G. Da Prato, L. Tubaro (Eds.), Stochastic Partial Differential Equations and Applications II. Proceedings, 1988. VI, 258 pages. 1989.

Vol. 1391: S. Cambanis, A. Weron (Eds.), Probability Theory on Vector Spaces IV. Proceedings, 1987. VIII, 424 pages. 1989.

Vol. 1392: R. Silhol, Real Algebraic Surfaces. X, 215 pages. 1989.

Vol. 1393: N. Bouleau, D. Feyel, F. Hirsch, G. Mokobodzki (Eds.), Séminaire de Théorie du Potentiel Paris, No. 9. Proceedings. VI, 265 pages. 1989.

Vol. 1394: T.L. Gill, W.W. Zachary (Eds.), Nonlinear Semigroups, Partial Differential Equations and Attractors. Proceedings, 1987. IX, 233 pages. 1989.

Vol. 1395: K. Alladi (Ed.), Number Theory, Madras 1987. Proceedings. VII, 234 pages. 1989.

Vol. 1396: L. Accardi, W. von Waldenfels (Eds.), Quantum Probability and Applications IV. Proceedings, 1987. VI, 355 pages. 1989.

Vol. 1397: P.R. Turner (Ed.), Numerical Analysis and Parallel Processing. Seminar, 1987. VI, 264 pages. 1989.

Vol. 1398: A.C. Kim, B.H. Neumann (Eds.), Groups – Korea 1988. Proceedings. V, 189 pages. 1989.

Vol. 1399: W.-P. Barth, H. Lange (Eds.), Arithmetic of Complex Manifolds. Proceedings, 1988. V, 171 pages. 1989.

Vol. 1400: U. Jannsen, Mixed Motives and Algebraic K-Theory. XIII, 246 pages. 1990.

Vol. 1401: J. Steprans, S. Watson (Eds.), Set Theory and its Applications. Proceedings, 1987. V, 227 pages. 1989.

Vol. 1402: C. Carasso, P. Charrier, B. Hanouzet, J.-L. Joly (Eds.), Nonlinear Hyperbolic Problems. Proceedings, 1988. V, 249 pages. 1989.

Vol. 1403: B. Simeone (Ed.), Combinatorial Optimization. Seminar, 1986. V, 314 pages. 1989.

Vol. 1404: M.-P. Malliavin (Ed.), Séminaire d´Algèbre Paul Dubreil et Marie-Paul Malliavin. Proceedings, 1987–1988. IV, 410 pages. 1989.

Vol. 1405: S. Dolecki (Ed.), Optimization. Proceedings, 1988. V, 223 pages. 1989. Vol. 1406: L. Jacobsen (Ed.), Analytic Theory of Continued Fractions III. Proceedings, 1988. VI, 142 pages. 1989.

Vol. 1407: W. Pohlers, Proof Theory. VI, 213 pages. 1989.

Vol. 1408: W. Lück, Transformation Groups and Algebraic K-Theory. XII, 443 pages. 1989.

Vol. 1409: E. Hairer, Ch. Lubich, M. Roche. The Numerical Solution of Differential-Algebraic Systems by Runge-Kutta Methods. VII, 139 pages. 1989.

Vol. 1410: F.J. Carreras, O. Gil-Medrano, A.M. Naveira (Eds.), Differential Geometry. Proceedings, 1988. V, 308 pages. 1989.

Vol. 1411: B. Jiang (Ed.), Topological Fixed Point Theory and Applications. Proceedings. 1988. VI, 203 pages. 1989.

Vol. 1412: V.V. Kalashnikov, V.M. Zolotarev (Eds.), Stability Problems for Stochastic Models. Proceedings, 1987. X, 380 pages. 1989.

Vol. 1413: S. Wright, Uniqueness of the Injective III Factor. III, 108 pages. 1989.

Vol. 1414: E. Ramirez de Arellano (Ed.), Algebraic Geometry and Complex Analysis. Proceedings, 1987. VI, 180 pages. 1989.

Vol. 1415: M. Langevin, M. Waldschmidt (Eds.), Cinquante Ans de Polynômes. Fifty Years of Polynomials. Proceedings, 1988. IX, 235 pages.1990.

Vol. 1416: C. Albert (Ed.), Géométrie Symplectique et Mécanique. Proceedings, 1988. V, 289 pages. 1990.

Vol. 1417: A.J. Sommese, A. Biancofiore, E.L. Livorni (Eds.), Algebraic Geometry. Proceedings, 1988. V, 320 pages. 1990.

Vol. 1418: M. Mimura (Ed.), Homotopy Theory and Related Topics. Proceedings, 1988. V, 241 pages. 1990.

Vol. 1419: P.S. Bullen, P.Y. Lee, J.L. Mawhin, P. Muldowney, W.F. Pfeffer (Eds.), New Integrals. Proceedings, 1988. V, 202 pages. 1990.

Vol. 1420: M. Galbiati, A. Tognoli (Eds.), Real Analytic Geometry. Proceedings, 1988. IV, 366 pages. 1990.

Vol. 1421: H.A. Biagioni, A Nonlinear Theory of Generalized Functions, XII, 214 pages. 1990.

Vol. 1422: V. Villani (Ed.), Complex Geometry and Analysis. Proceedings, 1988. V, 109 pages. 1990.

Vol. 1423: S.O. Kochman, Stable Homotopy Groups of Spheres: A Computer-Assisted Approach. VIII, 330 pages. 1990.

Vol. 1424: F.E. Burstall, J.H. Rawnsley, Twistor Theory for Riemannian Symmetric Spaces. III, 112 pages. 1990.

Vol. 1425: R.A. Piccinini (Ed.), Groups of Self-Equivalences and Related Topics. Proceedings, 1988. V, 214 pages. 1990.

Vol. 1426: J. Azéma, P.A. Meyer, M. Yor (Eds.), Séminaire de Probabilités XXIV, 1988/89. V, 490 pages. 1990.

Vol. 1427: A. Ancona, D. Geman, N. Ikeda, École d'Eté de Probabilités de Saint Flour

XVIII, 1988. Ed.: P.L. Hennequin. VII, 330 pages. 1990.

Vol. 1428: K. Erdmann, Blocks of Tame Representation Type and Related Algebras. XV. 312 pages. 1990.

Vol. 1429: S. Homer, A. Nerode, R.A. Platek, G.E. Sacks, A. Scedrov, Logic and Computer Science. Seminar, 1988. Editor: P. Odifreddi. V, 162 pages. 1990.

Vol. 1430: W. Bruns, A. Simis (Eds.), Commutative Algebra. Proceedings. 1988. V, 160 pages. 1990.

Vol. 1431: J.G. Heywood, K. Masuda, R. Rautmann, V.A. Solonnikov (Eds.), The Navier-Stokes Equations – Theory and Numerical Methods. Proceedings, 1988. VII, 238 pages. 1990.

Vol. 1432: K. Ambos-Spies, G.H. Müller, G.E. Sacks (Eds.), Recursion Theory Week. Proceedings, 1989. VI, 393 pages. 1990.

Vol. 1433: S. Lang, W. Cherry, Topics in Nevanlinna Theory. II, 174 pages.1990.

Vol. 1434: K. Nagasaka, E. Fouvry (Eds.), Analytic Number Theory. Proceedings, 1988. VI, 218 pages. 1990.

Vol. 1435: St. Ruscheweyh, E.B. Saff, L.C. Salinas, R.S. Varga (Eds.), Computational Methods and Function Theory. Proceedings, 1989. VI, 211 pages. 1990.

Vol. 1436: S. Xambó-Descamps (Ed.), Enumerative Geometry. Proceedings, 1987. V, 303 pages. 1990.

Vol. 1437: H. Inassaridze (Ed.), K-theory and Homological Algebra. Seminar, 1987–88. V, 313 pages. 1990.

Vol. 1438: P.G. Lemarié (Ed.) Les Ondelettes en 1989. Seminar. IV, 212 pages. 1990.

Vol. 1439: E. Bujalance, J.J. Etayo, J.M. Gamboa, G. Gromadzki. Automorphism Groups of Compact Bordered Klein Surfaces: A Combinatorial Approach. XIII, 201 pages. 1990.

Vol. 1440: P. Latiolais (Ed.), Topology and Combinatorial Groups Theory. Seminar, 1985–1988. VI, 207 pages. 1990.

Vol. 1441: M. Coornaert, T. Delzant, A. Papadopoulos. Géométrie et théorie des groupes. X, 165 pages. 1990.

Vol. 1442: L. Accardi, M. von Waldenfels (Eds.), Quantum Probability and Applications V. Proceedings, 1988. VI, 413 pages. 1990.

Vol. 1443: K.H. Dovermann, R. Schultz, Equivariant Surgery Theories and Their Periodicity Properties. VI, 227 pages. 1990.

Vol. 1444: H. Korezlioglu, A.S. Ustunel (Eds.), Stochastic Analysis and Related Topics VI. Proceedings, 1988. V, 268 pages. 1990.

Vol. 1445: F. Schulz, Regularity Theory for Quasilinear Elliptic Systems and – Monge Ampère Equations in Two Dimensions. XV, 123 pages. 1990.

Vol. 1446: Methods of Nonconvex Analysis. Seminar, 1989. Editor: A. Cellina. V, 206 pages. 1990.

Vol. 1447: J.-G. Labesse, J. Schwermer (Eds), Cohomology of Arithmetic Groups and Automorphic Forms. Proceedings, 1989. V, 358 pages. 1990.

Vol. 1448: S.K. Jain, S.R. López-Permouth (Eds.), Non-Commutative Ring Theory. Proceedings, 1989. V, 166 pages. 1990.

Vol. 1449: W. Odyniec, G. Lewicki, Minimal Projections in Banach Spaces. VIII, 168 pages. 1990.

Vol. 1450: H. Fujita, T. Ikebe, S.T. Kuroda (Eds.), Functional-Analytic Methods for Partial Differential Equations. Proceedings, 1989. VII, 252 pages. 1990.

Vol. 1451: L. Alvarez-Gaumé, E. Arbarello, C. De Concini, N.J. Hitchin, Global Geometry and Mathematical Physics. Montecatini Terme 1988. Seminar. Editors: M. Francaviglia, F. Gherardelli. IX, 197 pages. 1990.

Vol. 1452: E. Hlawka, R.F. Tichy (Eds.), Number-Theoretic Analysis. Seminar, 1988–89. V, 220 pages. 1990.

Vol. 1453: Yu.G. Borisovich, Yu.E. Gliklikh (Eds.), Global Analysis – Studies and Applications IV. V, 320 pages. 1990.

Vol. 1454: F. Baldassari, S. Bosch, B. Dwork (Eds.), p-adic Analysis. Proceedings, 1989. V, 382 pages. 1990.

Vol. 1455: J.-P. Françoise, R. Roussarie (Eds.), Bifurcations of Planar Vector Fields. Proceedings, 1989. VI, 396 pages. 1990.

Vol. 1456: L.G. Kovács (Ed.), Groups – Canberra 1989. Proceedings. XII, 198 pages. 1990.

Vol. 1457: O. Axelsson, L.Yu. Kolotilina (Eds.), Preconditioned Conjugate Gradient Methods. Proceedings, 1989. V, 196 pages. 1990.

Vol. 1458: R. Schaaf, Global Solution Branches of Two Point Boundary Value Problems. XIX, 141 pages. 1990.

Vol. 1459: D. Tiba, Optimal Control of Nonsmooth Distributed Parameter Systems. VII, 159 pages. 1990.

Vol. 1460: G. Toscani, V. Boffi, S. Rionero (Eds.), Mathematical Aspects of Fluid Plasma Dynamics. Proceedings, 1988. V, 221 pages. 1991.

Vol. 1461: R. Gorenflo, S. Vessella, Abel Integral Equations. VII, 215 pages. 1991.

Vol. 1462: D. Mond, J. Montaldi (Eds.), Singularity Theory and its Applications. Warwick 1989, Part I. VIII, 405 pages. 1991.

Vol. 1463: R. Roberts, I. Stewart (Eds.), Singularity Theory and its Applications. Warwick 1989, Part II. VIII, 322 pages. 1991.

Vol. 1465: G. David, Wavelets and Singular Integrals on Curves and Surfaces. X, 107 pages. 1991.

Vol. 1466: W. Banszczyk, Additive Subgroups of Topological Vector Spaces. VII, 178 pages. 1991.

Vol. 1467: W. M. Schmidt, Diophantine Approximations and Equations. VIII, 217 pages. 1991.

Vol. 1468: J. Noguchi, T. Ohsawa (Eds.), Prospects in Complex Geometry. Proceedings, 1989. VII, 421 pages. 1991.

Vol. 1469: J. Lindenstrauss, V. D. Milman (Eds.), Geometric Aspects of Functional Analysis. Seminar 1989-90. XI, 191 pages. 1991.